Impact of Farmland Abandonment on Water Resources and Soil Conservation

Impact of Farmland Abandonment on Water Resources and Soil Conservation

Special Issue Editors

Noemí Lana-Renault

Estela Nadal-Romero

Erik Cammeraat

José Ángel Llorente

MDPI • Basel • Beijing • Wuhan • Barcelona • Belgrade • Manchester • Tokyo • Cluj • Tianjin

Special Issue Editors

Noemí Lana-Renault
Universidad de La Rioja
Spain

José Ángel Llorente
Universidad de La Rioja
Spain

Estela Nadal-Romero
Instituto Pirenaico de Ecología
(IPE-CSIC)
Spain

Erik Cammeraat
Universiteit van Amsterdam
The Netherlands

Editorial Office
MDPI
St. Alban-Anlage 66
4052 Basel, Switzerland

This is a reprint of articles from the Special Issue published online in the open access journal *Water* (ISSN 2073-4441) (available at: https://www.mdpi.com/journal/water/special_issues/Farmland_ Abandonment).

For citation purposes, cite each article independently as indicated on the article page online and as indicated below:

LastName, A.A.; LastName, B.B.; LastName, C.C. Article Title. *Journal Name* **Year**, *Article Number*, Page Range.

ISBN 978-3-03936-611-8 (Hbk)
ISBN 978-3-03936-612-5 (PDF)

Contents

About the Special Issue Editors

Noemí Lana-Renault, Ph.D, is a physical geographer who has worked in several universities and research institutions in Spain and The Netherlands. Currently, she is Associate Professor at the University of La Rioja in Spain. Her research has focused on the impact of farmland abandonment on water resources and soil conservation, especially in the Mediterranean region. Most of her work is based on the analysis and interpretation of field data collected in small experimental catchments with different land uses and land covers.

Estela Nadal-Romero, Ph.D, is a physical geographer who has worked in several universities and research institutions in Spain, Belgium, and The Netherlands. Currently, she is a tenured scientist in the Pyrenean Institute of Ecology (IPE-CSIC). Her research is devoted to the integration of interdisciplinary knowledge derived from the geomorphological, hydrological, climatology, soil science, and ecological disciplines for the study of Mediterranean mountain areas.

Erik Cammeraat, Ph.D, works as an ssociate orofessor within the Ecosystem and Landscape Dynamics research department of the Institute of Biodiversity and Ecosystem Dynamics (IBED) of the University of Amsterdam (The Netherlands). His research interests are related to geomorphological, hydrological, soil, and land degradation processes, and specially focus on the development of emergent landscape properties as a result from the interaction between different physical, chemical, or biological processes; regreening of degraded environments; the role and fate of organic carbon in erosion and sedimentation; and planetary science, using terrestrial processes to understand analogue situations under non-terrestrial gravity and atmospheric conditions

José Ángel Llorente, Ph.D, is a physical geographer who works as a lecturer at the University of La Rioja. His research focuses on soil erosion in vineyards and land abandonment in agricultural terraces. He also woks on aspects related with the didactics of geography.

Editorial

Critical Environmental Issues Confirm the Relevance of Abandoned Agricultural Land

Noemí Lana-Renault [1,*], Estela Nadal-Romero [2], Erik Cammeraat [3] and José Ángel Llorente [1]

[1] Departamento de Ciencias Humanas, Universidad de La Rioja, 26004 Logrono, Spain; jose-angel.llorente@unirioja.es
[2] Instituto Pirenaico de Ecología, IPE-CSIC, 50059 Zaragoza, Spain; estelanr@ipe.csic.es
[3] Institute for Biodiversity and Ecosystem Dynamics, Universiteit van Amsterdam, 1098XH Amsterdam, The Netherlands; L.H.Cammeraat@uva.nl
* Correspondence: noemi-solange.lana-renault@unirioja.es; Tel.: +34-941-299-553

Received: 15 March 2020; Accepted: 8 April 2020; Published: 14 April 2020

Abstract: Large areas worldwide have been affected by farmland abandonment and subsequent plant colonization with significant environmental consequences. Although the process of farmland abandonment has slowed down, vegetation recovery in abandoned lands is far from complete. In addition, agricultural areas and pasture lands with low-intensity activities could be abandoned in the near future. In this foreword, we review current knowledge of the impacts of farmland abandonment on water resources and soil conservation, and we highlight the open questions that still persist, in particular regarding terraced landscapes, afforested areas, abandonment of woody crops, traditional irrigated fields, solute yields, long-term trends in the response of abandoned areas, and the management of abandoned farmland. This Special Issue includes seven contributions that illustrate recent research into the hydrological, geomorphological, and edaphological consequences of farmland abandonment.

Keywords: farmland abandonment; land-use change; revegetation; hydrological response; soil erosion

---

1. Introduction

Farmland abandonment is a major land-use change in many rural territories, particularly in temperate, developed regions [1–3]. It usually affects less productive land that is less suitable for industrial-scale production [4] and includes both crops and pastures. In some instances, physical constraints (e.g., poor soils, scarce water resources, intense land degradation) have also caused the abandonment of agricultural land [5]. Finally, farmland abandonment can be induced by national or supranational policies, which regulate markets for particular products to the detriment of others [6,7].

Campbell et al. (2008) [2] estimated that 385–472 million hectares of farmland was abandoned worldwide between 1700 and 2000. 99% of this land abandonment occurred during the last 100 years of this period. It affected all parts of the world, but was more intense in the United States, Europe, and Australia. One of the first countries to experience farmland abandonment was the United States [1]. In the middle of the 19th century the northeastern regions started to be abandoned due to competition from agriculture in the Midwest and the Great Plains [8]. Waisanen and Bliss [9] estimated that around 75% of the agricultural land in these regions had been abandoned between 1880 and 1997. In Europe, farmland abandonment occurred mainly in the 20th century, more intensely after the 1950s [1]. Some regions, however, were already abandoned at the end of the 19th century. In Switzerland, farmland abandonment has been observed for more than 150 years [10]. In the Hérault region in France [11], abandonment of croplands at the end of the 19th century was detected. Fuchs et al. (2012) [12] suggested that in Europe cropland decreased by almost 19% between 1950 and

2010 and semi-natural grasslands by almost 6%. In some mountain regions, however the process of land abandonment has been very intense, affecting for instance more than 80% of cultivated land in the Spanish Pyrenees [13] and around 70% in the eastern Alps [14]. Recent abandonment has been observed elsewhere: Eastern Europe, China, Australia and, to a lesser extent, Canada, South America (e.g., Argentina, Brazil), northern Africa and India, have seen a decrease in farming since the end of the 20th century [1,2]. In the Loess Plateau (China), the Grain-for-Green programs was launched in 1999 by the Government in order to control intense soil degradation and has converted large areas of arable land into artificial forests or into areas left to spontaneous plant colonization [15], with a total restored area estimated at almost 5 million hectares [16]. Australia has one of the highest levels of pasture abandonment, which started in the 1970s [2]. In Canada, the Lower Inventory for Tomorrow of 1970 promoted the withdrawal of cultivated land from production by offering financial compensation to farmers [17]. In Europe, withdrawal of agricultural land has been induced by the Common Agricultural Policy (CAP) since the 1990s, mainly in plains and piedmonts, in order to reduce food surpluses and to limit the costs of agricultural subsidies [6]. In Eastern European countries, the extensive abandonment of agricultural land occurred following the fall of the communist regimes in the 1990s. The subsequent agrarian reforms and the change to a market-oriented economy caused the collapse of many collective farms and the abandonment of agricultural practices [18]. For instance, Nikodemus et al. (2005) [19] showed that in Latvia land abandonment was extensive, with 50% of croplands no longer cultivated by 1999.

The process of farmland abandonment has slowed down in most developed countries [3]. The forecast surface subject to abandonment is highly variable, depending on the land-use model employed and the scenarios considered, but there is general agreement that it will particularly affect marginal areas with low-intensity activities [20,21]. Rural areas in developing countries might undergo a similar process in the near future, with agricultural intensification in the more productive areas and a decline in farming in the less productive ones. However, this trend is very uncertain due to the great many economical, demographical and political factors that affect farmland abandonment [20]. For instance, some relatively steep areas in Mediterranean piedmonts have been affected by increasing pressure for re-cultivation due to favourable market conditions, e.g., vineyards and almond and olive tree orchards, resulting in severe erosion problems [22].

In most cases, the abandonment of agricultural land leads to a process of natural vegetation recovery [23] (Figures 1 and 2). Colonization by plants, also known as secondary succession, is very complex in abandoned fields. It depends on both natural and human-induced factors, including climate, topography, soil conditions, the distance and floristic composition of bordering vegetation, the age of abandonment and management following the end of farming, particularly livestock grazing and the occurrence of fires [24–26]. Research carried out worldwide has shown that vegetation recovery in abandoned lands has significant implications for landscape, water resources, soil erosion and biodiversity [4,5,17,27]. This Special Issue includes seven contributions that focus on the hydrological and geomorphological consequences (soil hydraulic properties, runoff, soil erosion, solute export) of farmland abandonment in different environments, including terraced landscapes, afforested areas, abandoned land at different stages of succession and recent abandonment in irrigated land and steep vineyards. In this foreword, we review briefly the main impacts of farmland abandonment on water resources and soil conservation and identify the main questions that still need to be addressed by scientists and land managers. Our main purpose is to highlight the environmental relevance of abandoned agricultural land and to show that its management is a challenge that requires immediate responses.

Figure 1. Plant colonization in abandoned fields in a wet environment (Arnás catchment, Central Spanish Pyrenees). Forest stands occupy the gentle slopes in the forefront whereas the steep slopes in the background are colonized by shrubs. Photo by Jérôme Latron.

Figure 2. Abandoned terraces in the Upper Guadalentin basin in Murcia (southeast Spain). Land abandonment and slow revegetation produce soil erosion processes. Photo by Estela Nadal-Romero.

2. Brief Review of the Main Impacts of Farmland Abandonment on Water Resources and Soil Conservation

Vegetation expansion alters the water cycle and the partitioning of precipitation between evapotranspiration, runoff and groundwater flows. Thus, water yields usually decrease following revegetation due to increased rainfall interception and transpiration by forests and shrubs [28].

However, the hydrological impact of these processes varies greatly as it depends on several factors [29], such as climate conditions, the extent and spatial distribution of land-cover disturbance, the characteristics and depth of the soils, and the type, age or physiology of the vegetation.

The hydrological consequences of vegetation recovery on formerly cultivated land have been examined at different spatial scales. Several studies have reported declining river discharges and negative trends in highflows as shrubs and forest expand at the headwaters (e.g. [30–33]). At the small-catchment scale, studies worldwide have demonstrated that vegetation established after land abandonment noticeably reduces runoff coefficients and is a major factor affecting flood control (e.g. [34–38]). Some of these studies have shown that revegetation also affects flood hydrograph characteristics, with lower peakflows, slower response times and slower recession limbs in forested catchments [39,40]. These results have been supported by research carried out at a more detailed scale. For instance, Nadal-Romero et al. (2013) [41] reported very low or no overland flow on abandoned plots, especially on those covered by dense vegetation, when compared with plots covered by cereal crops or fallow land. In arid and semi-arid environments, however, the effect of land abandonment may be the opposite, as land abandonment may cause the formation of soil surface crusts that reduce soil infiltration and favor overland flow [42,43]. An increase in vegetation cover also reduces soil water content and aquifer recharge due to rainfall interception and high transpiration rates by trees; this effect is much stronger under dry conditions or in dry environments [15,44]. Finally, several studies [45,46] have shown that the expansion of forest may affect snow accumulation and distribution, as trees reduce beneath-canopy snow accumulation and alter snow melting rates, mainly due to interception and subsequent sublimation processes.

Land abandonment also has significant consequences for soil conservation and soil erosion [17]. Revegetation following land abandonment tends to improve soil properties in the long term [47], usually showing increased soil organic matter and soil fertility [48], greater aggregate stability [49], and higher infiltration capacity, water-holding capacity and hydraulic conductivity [50]. However, the evolution of these processes is conditioned by topography, soil quality prior to abandonment and land management after abandonment, especially the occurrence of fires [48].

Studies worldwide have demonstrated that the expansion of vegetation cover decreases soil erosion due to the protective role of vegetation against rainfall splash and reduces surface runoff. In abandoned arable land, soil erosion can be significant during the first stages of plant colonization, as shown by Ruiz-Flaño et al. (1992) [51] and by Cerdá et al. (2018) [52] for the Western Mediterranean region, but it decreases over time as vegetation cover becomes denser. In some cases, signs of severe erosion such as undermining of shrubs, rills or development of stone pavement have been detected in the oldest fields [51,53]. This is because these fields were normally located in the worst positions (e.g., steep slopes, stony soils) and were subject to recurrent burning for shifting agriculture [41], factors that constrained the establishment of dense vegetation cover. In mountain pasture lands, abandonment may initially trigger shallow landslides due to a change in plant communities, as shown by Tasser et al. (2003) [54] in several sub-alpine and alpine meadows in Europe. When precipitation is scarce and irregular, such as in arid and semi-arid regions, plant colonization is difficult and geomorphic activity is more intense. Moreover, the formation of soil surface crusts that promote overland flow enhances soil erosion, resulting in severe sheet wash erosion [55] and the development of rills and gullies [56]. In these environments, particular soil characteristics such as the presence of marls may accelerate soil erosion processes (i.e. piping) due to the dispersion of clay minerals [56]. As a consequence of soil erosion, abandoned fields may be subject to a significant loss of nutrients and organic matter, and a decline in soil quality [17]. Finally, extensive revegetation on formerly cultivated lands has favored the occurrence and propagation of wildfires (e.g. [57,58]), which usually cause intense soil erosion in the first years following the fire due to greater overland flow [59].

Likewise, the general expansion of vegetation implies a shrinkage of sediment sources [60] and a decrease in sediment supply from the hillslopes to the channels [61,62]. In semi-arid areas, although runoff and soil erosion can be locally high after farmland abandonment, vegetation tends to develop in patches that act as water and sediment sinks, thus reducing sediment delivery to the stream [42,63]. As a result, the rate of sedimentation levels of reservoirs, rivers and coastal areas has declined [64,65] and the morphology of rivers has changed, with the narrowing and incision of alluvial plains [66,67] and subsequent environmental implications [68].

The effects of farmland abandonment on terraced landscapes are manifold due to the great complexity of the hydrological and geomorphological processes that affect these man-made constructions [69,70]. Farmland abandonment often leads to the collapse of terrace structures that are no longer maintained. Soil saturation in the inner part of the terraces, usually caused by the inefficiency of the drainage systems following abandonment [71], encourages wall instability and the occurrence of mass movements. The failure of terrace risers results in scars that are frequently affected by gullying or livestock trampling. Modeling exercises have shown that the formation of gullies reactivates the original drainage network, so favoring hillslope-channel connections and increasing peakflow discharges and sediment yield (e.g. [72,73]). On very steep slopes, debris slips and cascade landslides have been observed under intense rainfall [74]. The development of vegetation cover tends to reduce geomorphic activity on terraces (especially surface erosion processes), but often fails to prevent terrace collapse [70]. In semi-arid fields with marl lithology, terrace abandonment enhances intense surface erosion and favors piping, which may lead to deep gully incisions [75].

3. Open Questions

Although the consequences of farmland abandonment on water resources and soil conservation have been extensively studied, unresolved questions persist and new scientific challenges have emerged. For instance, traditional woody crops such as vineyards and fruit orchards in both rainfed and irrigated areas have been abandoned in recent decades or could be abandoned in the near future [21,76]. Literature about the consequences of the abandonment of such traditional systems on soil conservation and hydrology is still scarce [76,77].

There are still many open questions concerning the abandonment of terraced landscapes: what is the frequency of terrace collapse and what are the temporal drivers [78]? To what extent and under which conditions can the original drainage network be reinstalled? What are the downstream effects of the degradation of terraced slopes? Some authors support the conservation and rehabilitation of terraced structures because of their environmental, productive and aesthetic functions [79,80], but a thorough assessment of such efforts is still lacking.

The published literature provides substantial information on how farmland abandonment affects soil properties and soil erosion. However, limited information is available on how the abandonment of agricultural land affects surface water quality and solute export [81]. This is important for understanding changes in soil fertility in abandoned lands and the risk of soil degradation. Similarly, more work has to be done on how much soils in abandoned fields are affected by water repellency [82], a phenomenon that decreases infiltration capacity and may explain the high variability of runoff responses in abandoned lands.

Although farmland abandonment is predicted to slow down, vegetation recovery in abandoned lands is far from being complete. There are still large areas covered with shrubs that should evolve into forest stands and large areas of high-altitude pastures that have been colonized by shrubs and trees over the last few decades (Figure 3), enhanced by livestock decline (particularly sheep and goats), a warming climate and the rise of atmospheric CO_2 [83]. This means that vegetation in former agricultural and pasture land will continue to expand in the coming decades, with increasing impacts on water resources and soil conservation. Moreover, the evolution of vegetation in fields abandoned for very long periods may give rise to new processes affecting the hydrological and geomorphological dynamics of these areas. For instance, as well as improving soil quality, the higher input of organic matter due to vegetation favors soil development [84] and a consequent increase in long-term soil water storage. There is evidence that transpiration of old trees is lower than that of young individuals [85], suggesting that the impact of revegetation on runoff may stabilize or even decrease over time. In some cases, after decades of abandonment, degradation of shrub cover due to senescence has been observed, indicating a possible increase in runoff and sediment yield in the long term [41]. All these questions highlight the need for further research based on long-term data series in order to detect trends and changes in the system response.

Figure 3. Recent colonization by *P. Uncinata* in the subalpine belt in the Spanish Pyrenees (Las Blancas), as a consequence of a decline in livestock grazing (particularly sheep). The taller trees are 20 years old and the small trees are 3–4 years old. Photo by José M. García-Ruiz.

Extensive afforestation programs have been established in formerly cultivated areas by national forest services in order to improve the use of abandoned land as a resource and to control hydrological and soil-erosion processes [86]. At present, though the time elapsed since the first plantations provides sufficient perspective to assess their hydrological and geomorphological efficiency, there have been few studies at the catchment scale (e.g. [62,87]).

One of the topics that currently arouses special interest is the management of abandoned land [88]. Till now, as most abandoned areas have been considered marginal in economic terms, they have lacked any management intervention, leading to a process of "rewilding" or landscape naturalization [89]. As shown above, the recovery of vegetation may have important environmental benefits such as soil conservation, moderation of the hydrological response (in the sense that peakflows are lower and are delayed) and improved water quality. It also increases carbon sequestration [90,91]. For some authors, it is the best option for nature conservation and biodiversity [92], although there are differing opinions on the latter. Many authors argue that the effects of rewilding on biodiversity can be both positive and negative, depending on the species considered [93]. Finally, the regeneration of forest may also enhance the recreational value of the landscape [94]. However, land abandonment can also have negative impacts: as seen above, expansion of shrubs and forests decreases runoff and reduces river discharges in many cases; the presence of dense, continuous forests favors the occurrence of large wildfires; when vegetation succession is very slow or is interrupted, intense soil degradation may occur [17]. Additionally, landscape naturalization is often seen as a significant loss of traditional landscapes and development opportunities for the local population [95]. Thus, for some authors, active management of abandoned land is necessary in order to improve the ecosystem services these areas provide to society [88]. They consider that forest intervention, including the recovery of degraded ecosystems and the control of plant succession by light human activity (e.g., shrub clearing for extensive stockbreeding), will benefit biodiversity, land productivity, water resources, soil conservation and wildfire risk [96]. At present, there is an intense debate regarding rewilding *versus* forest intervention and there is still no clear consensus about the optimal strategy for the management of old agricultural areas.

4. Final Remarks

Abandoned agricultural areas deserve special attention because of their influence on the provision of natural resources such as water and on soil conservation. However, benefits depend greatly on how these areas are managed, which, at present, is a controversial topic with conflicting positions. In the current context of degradation of natural resources and climate change projections, there is an urgent need for scientists to provide knowledge to help decide on the best way to manage former agricultural land. This would optimize the environmental services they can supply to society.

We would like to conclude this foreword by acknowledging the efforts of the authors and reviewers of the articles presented in this Special Issue. We hope very much that these are useful and relevant to the readers of Water.

Author Contributions: All authors led the development of the Special Issue and contributed to the writing of this foreword. All authors have read and agreed to the published version of the manuscript.

Funding: Support for this research was provided by the projects ESPAS (CGL2015-65569-R, funded by the Spanish Ministry of Economy and Competitiveness) and LIFE-MIDMACC (LIFE18 CCA/ES/001099, funded by the European Commission).

Conflicts of Interest: The authors declare no conflict of interest.

References

1. Ramankutty, N.; Foley, J.A. Estimating historical changes in global land cover. *Glob. Biogeochem. Cycles* **1999**, *13*, 997–1027. [CrossRef]
2. Campbell, J.E.; Lobell, D.B.; Genova, R.C.; Field, C.B. The global potential of bioenergy on abandoned agriculture lands. *Environ. Sci. Technol.* **2008**, *42*, 5791–5794. [CrossRef] [PubMed]
3. Ellis, E.C.; Kaplan, J.O.; Fuller, D.Q.; Vavrus, S.; Goldewijk, K.K.; Verburg, P.H. Used planet: A global history. *Proc. Natl. Acad. Sci. USA* **2013**, *110*, 7978–7985. [CrossRef] [PubMed]
4. Rey-Benayas, J.M.; Martins, A.; Nicolau, J.M.; Schulz, J.J. Abandonment of agricultural land: An overview of drivers and consequences. *CAB Rev. Perspect. Agric. Vet. Sci. Nutr. Nat. Resour.* **2007**, *2*, 1–14. [CrossRef]
5. García-Ruiz, J.M.; Lana-Renault, N. Hydrological and erosive consequences of farmland abandonment in Europe, with special reference to the mediterranean region—A review. *Agric. Ecosyst. Environ.* **2011**, *140*, 317–338. [CrossRef]
6. Lasanta, T.; Arnáez, J.; Pascual, N.; Ruiz-Flaño, P.; Errea, M.P.; Lana-Renault, N. Space-time process and drivers of land abandonment in Europe. *Catena* **2017**, *149*, 810–823. [CrossRef]
7. Van Leeuwen, C.C.E.; Cammeraat, E.L.H.; de Vente, J.; Boix-Fayos, C. The evolution of soil conservation policies targeting land abandonment and soil erosion in Spain: A review. *Land Use Policy* **2019**, *83*, 174–186. [CrossRef]
8. McGrory Klyza, C. *Wilderness Comes Home: Rewilding the Northeast*; Middlebury College Press: Middlebury, VT, USA, 2001.
9. Waisanen, P.J.; Bliss, N.B. Changes in population and agricultural land in conterminous United States counties, 1790 to 1997. *Glob. Biogeochem. Cycles* **2002**, *16*, 1–19. [CrossRef]

10. Gellrich, M.; Zimmermann, N.E. Investigating the regional-scale pattern of agricultural land abandonment in the Swiss mountains: A spatial statistical modelling approach. *Landsc. Urban Plan.* **2007**, *79*, 65–76. [CrossRef]

11. Debussche, M.; Lepart, J.; Dervieux, A. mediterranean landscape changes: Evidence from old postcards. *Glob. Ecol. Biogeogr.* **1999**, *8*, 3–15. [CrossRef]

12. Fuchs, R.; Herold, M.; Verburg, P.H.; Clevers, J.G.P.W. A high-resolution and harmonized model approach for reconstructing and analysing historic land changes in Europe. *Biogeosciences* **2013**, *10*, 1543–1559. [CrossRef]

13. Lasanta, T. The process of desertion of cultivated areas in the central spanish Pyrenees. *Pirineos* **1988**, *132*, 15–36.

14. Tasser, E.; Walde, J.; Tappeiner, U.; Teutsch, A.; Noggler, W. Land-use changes and natural reforestation in the eastern central Alps. *Agric. Ecosyst. Environ.* **2007**, *118*, 115–129. [CrossRef]

15. Cao, S.; Chen, L.; Yu, X. Impact of China's Grain for Green Project on the landscape of vulnerable arid and semi-arid agricultural regions: A case study in northern shaanxi province. *J. Appl. Ecol.* **2009**, *46*, 536–543. [CrossRef]

16. Zhao, G.; Mu, X.; Wen, Z.; Wang, F.; Gao, P. Soil erosion, conservation, and eco-environment changes in the loess plateau of china. *Land Degrad. Dev.* **2013**, *24*, 499–510. [CrossRef]

17. Lasanta, T.; Arnáez, J.; Nadal-Romero, E. *Soil Degradation, Restoration and Management in Abandoned and Afforested Lands*; Elsevier: Amsterdam, The Netherlands, 2019; Volume 4, pp. 71–117.

18. Prishchepov, A.A.; Müller, D.; Dubinin, M.; Baumann, M.; Radeloff, V.C. Determinants of agricultural land abandonment in post-Soviet European Russia. *Land Use Policy* **2013**, *30*, 873–884. [CrossRef]

19. Nikodemus, O.; Bell, S.; Grine, I.; Liepiņš, I. The impact of economic, social and political factors on the landscape structure of the vidzeme uplands in Latvia. *Landsc. Urban Plan.* **2005**, *70*, 57–67. [CrossRef]

20. Keenleyside, C.; Tucker, G.M. *Farmland Abandonment in the Eu: An Assessment of Trends and Prospects*; Report prepared for WWF; Institute for European Environmental Policy: London, UK, 2010.

21. Malek, Ž.; Verburg, P.H.R.; Geijzendorffer, I.; Bondeau, A.; Cramer, W. Global change effects on land management in the mediterranean region. *Glob. Environ. Chang.* **2018**, *50*, 238–254. [CrossRef]

22. Martínez-Casasnovas, J.A.; Ramos, M.C. The cost of soil erosion in vineyard fields in the Penedès-Anoia region (NE Spain). *Catena* **2006**, *68*, 194–199. [CrossRef]

23. Sluiter, R.; De Jong, S.M. Spatial patterns of mediterranean land abandonment and related land cover transitions. *Landsc. Ecol.* **2007**, *22*, 559–576. [CrossRef]

24. Chauchard, S.; Carcaillet, C.; Guibal, F. Patterns of land-use abandonment control tree-recruitment and forest dynamics in mediterranean mountains. *Ecosystems* **2007**, *10*, 936–948. [CrossRef]

25. Peña-Angulo, D.; Khorchani, M.; Errea, P.; Lasanta, T.; Martínez-Arnáiz, M.; Nadal-Romero, E. Factors explaining the diversity of land cover in abandoned fields in a mediterranean mountain area. *Catena* **2019**, *181*, 104064. [CrossRef]

26. Gómez, D.; Aguirre, A.J.; Lizaur, X.; Lorda, M.; Remón, J.L. Evolution of argoma shrubland (*Ulex gallii* Planch.) after clearing and burning treatments in Sierra de Aralar and belate (Navarra). *Geogr. Res. Lett.* **2019**, *45*, 469–486.

27. Sitzia, T.; Semenzato, P.; Trentanovi, G. Natural reforestation is changing spatial patterns of rural mountain and hill landscapes: A global overview. *Ecol. Manag.* **2010**, *259*, 1354–1362. [CrossRef]

28. Bosch, J.M.; Hewlett, J.D. A review of catchment experiments to determine the effect of vegetation changes on water yield and evapotranspiration. *J. Hydrol.* **1982**, *55*, 3–23. [CrossRef]

29. Lana-Renault, N.; Morán-Tejeda, E.; Moreno-de-las-Heras, M.; Lorenzo-Lacruz, J.; López-Moreno, J.I. Land-use change and impacts. In *Water Resources in the Mediterranean Region*; Zribi, M., Brocca, L., Tramblay, Y., Molle, F., Eds.; Elsevier: Amsterdam, The Netherlands, 2000.

30. Gallart, F.; Llorens, P. Catchment management under environmental change: Impact of land cover change on water resources. *Water Int.* **2003**, *28*, 334–340. [CrossRef]

31. López-Moreno, J.I.; Beguería, S.; García-Ruiz, J.M. Trends in high flows in the central Spanish Pyrenees: Response to climatic factors or to land-use change? *Hydrol. Sci. J.* **2006**, *51*, 1039–1050. [CrossRef]

32. Morán-Tejeda, E.; Ceballos-Barbancho, A.; Llorente-Pinto, J.M. Hydrological response of mediterranean headwaters to climate oscillations and land-cover changes: The mountains of Duero river basin (central Spain). *Glob. Planet. Chang.* **2010**, *72*, 39–49. [CrossRef]

33. Martínez-Fernández, J.; Sánchez, N.; Herrero-Jiménez, C.M. Recent trends in rivers with near-natural flow regime: The case of the river headwaters in Spain. *Prog. Phys. Geogr.* **2013**, *37*, 685–700. [CrossRef]

34. Cosandey, C.; Andréassian, V.; Martin, C.; Didon-Lescot, J.F.; Lavabre, J.; Folton, N.; Mathys, N.; Richard, D. The hydrological impact of the mediterranean forest: A review of French research. *J. Hydrol.* **2005**, *301*, 235–249. [CrossRef]

35. García-Ruiz, J.M.; Regüés, D.; Alvera, B.; Lana-Renault, N.; Serrano-Muela, P.; Nadal-Romero, E.; Navas, A.; Latron, J.; Martí-Bono, C.; Arnáez, J. Flood generation and sediment transport in experimental catchments affected by land use changes in the central Pyrenees. *J. Hydrol.* **2008**, *356*, 245–260. [CrossRef]

36. Lana-Renault, N.; Nadal-Romero, E.; Serrano-Muela, M.P.; Alvera, B.; Sánchez-Navarrete, P.; Sanjuan, Y.; García-Ruiz, J.M. Comparative analysis of the response of various land covers to an exceptional rainfall event in the central spanish Pyrenees, October 2012. *Earth Surf. Process. Landf.* **2014**, *39*, 581–592. [CrossRef]

37. Rodríguez-Caballero, E.; Lázaro, R.; Cantón, Y.; Puigdefábregas, J.; Solé-Benet, A. Long-term hydrological monitoring in arid-semiarid almerÍa, SE Spain. What have we learned? *Geogr. Res. Lett.* **2018**, *44*, 581–600. [CrossRef]

38. Fortesa, J.; Latron, J.; García-Comendador, J.; Tomàs-Burguera, M.; Company, J.; Calsamiglia, A.; Estrany, J. Multiple temporal scales assessment in the hydrological response of small mediterranean-climate catchments. *Water* **2020**, *12*, 299. [CrossRef]

39. Burch, G.J.; Bath, R.K.; Moore, I.D.; O'Loughlin, E.M. Comparative hydrological behaviour of forested and cleared catchments in Southeastern Australia. *J. Hydrol.* **1987**, *90*, 19–42. [CrossRef]

40. Lana-Renault, N.; Latron, J.; Karssenberg, D.; Serrano-Muela, P.; Regüés, D.; Bierkens, M.F.P. Differences in stream flow in relation to changes in land cover: A comparative study in two sub-mediterranean mountain catchments. *J. Hydrol.* **2011**, *411*, 366–378. [CrossRef]

41. Nadal-Romero, E.; Lasanta, T.; García-Ruiz, J.M. Runoff and sediment yield from land under various uses in a mediterranean mountain area: Long-term results from an experimental station. *Earth Surf. Process. Landf.* **2013**, *38*, 346–355. [CrossRef]

42. Cammeraat, L.H.; Imeson, A.C. The evolution and significance of soil-vegetation patterns following land abandonment and fire in Spain. *Catena* **1999**, *37*, 107–127. [CrossRef]

43. Ries, J.B.; Langer, M. Runoff generation on abandoned fields in the central Ebro basin. Results from rainfall simulation experiments. *Cuad. Investig. Geográfica* **2001**, *27*, 61. [CrossRef]

44. Llorens, P.; Poyatos, R.; Latron, J.; Delgado, J.; Oliveras, I.; Gallart, F. A multi-year study of rainfall and soil water controls on Scots pine transpiration under mediterranean mountain conditions. *Hydrol. Process.* **2010**, *24*, 3053–3064. [CrossRef]

45. Lundquist, J.D.; Dickerson-Lange, S.E.; Lutz, J.A.; Cristea, N.C. Lower forest density enhances snow retention in regions with warmer winters: A global framework developed from plot-scale observations and modeling. *Water Resour. Res.* **2013**, *49*, 6356–6370. [CrossRef]

46. Revuelto, J.; López-Moreno, J.I.; Azorin-Molina, C.; Vicente-Serrano, S.M. Canopy influence on snow depth distribution in a pine stand determined from terrestrial laser data. *Water Resour. Res.* **2015**, *51*, 3476–3489. [CrossRef]

47. Van Hall, R.L.; Cammeraat, L.H.; Keesstra, S.D.; Zorn, M. Impact of secondary vegetation succession on soil quality in a humid mediterranean landscape. *Catena* **2017**, *149*, 836–843. [CrossRef]

48. Nadal, J.; Pèlachs, A.; Molina, D.; Soriano, J.M. Soil fertility evolution and landscape dynamics in a mediterranean area: A case study in the Sant Llorenç natural park (Barcelona, NE Spain). *Area* **2009**, *41*, 129–138. [CrossRef]

49. Ruecker, G.; Schad, P.; Alcubilla, M.M.; Ferrer, C. Natural regeneration of degraded soils and site changes on abandoned agricultural terraces in mediterranean Spain. *Land Degrad. Dev.* **1998**, *9*, 179–188. [CrossRef]

50. Martinez-Fernandez, J.; Lopez-Bermudez, F.; Martinez-Fernandez, J.; Romero-Diaz, A. Land use and soil-vegetation relationships in a mediterranean ecosystem: El Ardal, Murcia, Spain. *Catena* **1995**, *25*, 153–167. [CrossRef]

51. Ruiz-Flaño, P.; García-Ruiz, J.M.; Ortigosa, L. Geomorphological evolution of abandoned fields. A case study in the central Pyrenees. *Catena* **1992**, *19*, 301–308. [CrossRef]

52. Cerdà, A.; Rodrigo-Comino, J.; Novara, A.; Brevik, E.C.; Vaezi, A.R.; Pulido, M.; Giménez-Morera, A.; Keesstra, S.D. Long-term impact of rainfed agricultural land abandonment on soil erosion in the western mediterranean basin. *Prog. Phys. Geogr. Earth Environ.* **2018**, *42*, 202–219. [CrossRef]

53. Poesen, J.; De Luna, E.; Franca, A.; Nachtergaele, J.; Govers, G. Concentrated flow erosion rates as affected by rock fragment cover and initial soil moisture content. *Catena* **1999**, *36*, 315–329. [CrossRef]

54. Tasser, E.; Mader, M.; Tappeiner, U. Effects of land use in alpine grasslands on the probability of landslides. *Basic Appl. Ecol.* **2003**, *280*, 271–280. [CrossRef]

55. Sauer, T.; Ries, J.B. Vegetation cover and geomorphodynamics on abandoned fields in the central Ebro basin (Spain). *Geomorphology* **2008**, *102*, 267–277. [CrossRef]

56. Lesschen, J.P.; Kok, K.; Verburg, P.H.; Cammeraat, L.H. Identification of vulnerable areas for gully erosion under different scenarios of land abandonment in southeast Spain. *Catena* **2007**, *71*, 110–121. [CrossRef]

57. Pausas, J.G.; Fernández-Muñoz, S. Fire regime changes in the western mediterranean basin: From fuel-limited to drought-driven fire regime. *Clim. Chang.* **2012**, *110*, 215–226. [CrossRef]

58. Pastor, A.V.; Nunes, J.P.; Ciampalini, R.; Koopmans, M.; Baartman, J.; Huard, F.; Calheiros, T.; Le-Bissonnais, Y.; Keizer, J.J.; Raclot, D. Projecting future impacts of global change including fires on soil erosion to anticipate better land management in the forests of NW Portugal. *Water* **2019**, *11*, 2617. [CrossRef]

59. Shakesby, R.A.; Doerr, S.H. Wildfire as a hydrological and geomorphological agent. *Earth Sci. Rev.* **2006**, *74*, 269–307. [CrossRef]

60. Lana-Renault, N.; Regüés, D. Seasonal patterns of suspended sediment transport in an abandoned farmland catchment in the central spanish Pyrenees. *Earth Surf. Process. Landf.* **2009**, *34*, 1291–1301. [CrossRef]

61. Bakker, M.M.; Govers, G.; van Doorn, A.; Quetier, F.; Chouvardas, D.; Rounsevell, M. The response of soil erosion and sediment export to land-use change in four areas of Europe: The importance of landscape pattern. *Geomorphology* **2008**, *98*, 213–226. [CrossRef]

62. Piégay, H.; Walling, D.E.; Landon, N.; He, Q.; Liébault, F.; Petiot, R. Contemporary changes in sediment yield in an alpine mountain basin due to afforestation (the upper Drôme in France). *Catena* **2004**, *55*, 183–212. [CrossRef]

63. Puigdefábregas, J. The role of vegetation patterns in structuring runoff and sediment fluxes in drylands. *Earth Surf. Process. Landf.* **2005**, *30*, 133–147. [CrossRef]

64. Keesstra, S.D. Impact of natural reforestation on floodplain sedimentation in the Dragonja basin, SW Slovenia. *Earth Surf. Process. Landf.* **2007**, *32*, 49–65. [CrossRef]

65. López-Moreno, J.I.; Vicente-Serrano, S.M.; Moran-Tejeda, E.; Zabalza, J.; Lorenzo-Lacruz, J.; García-Ruiz, J.M. Impact of climate evolution and land use changes on water yield in the Ebro basin. *Hydrol. Earth Syst. Sci.* **2011**, *15*, 311–322. [CrossRef]

66. Beguería, S.; López-Moreno, J.I.; Gómez-Villar, A.; Rubio, V.; Lana-Renault, N.; García-Ruiz, J.M. Fluvial adjustments to soil erosion and plant cover changes in the central spanish Pyrenees. *Geogr. Ann. Ser. A Phys. Geogr.* **2006**, *88*, 177–186. [CrossRef]

67. Sanjuán, Y.; Gómez-Villar, A.; Nadal-Romero, E.; Álvarez-Martínez, J.; Arnáez, J.; Serrano-Muela, M.P.; Rubiales, J.M.; González-Sampériz, P.; García-Ruiz, J.M. Linking land cover changes in the Sub-Alpine and montane belts to changes in a Torrential river. *Land Degrad. Dev.* **2016**, *27*, 179–189. [CrossRef]

68. Halifa-Marín, A.; Pérez-Cutillas, P.; Almagro, M.; Boix-Fayos, C. Presión antrópica sobre cuencas de drenaje en ecosistemas frágiles: Variaciones en las existencias (stock) de carbono orgánico asociadas a cambios morfológicos fluviales. *Cuad. Investig. Geográfica* **2019**, *45*, 245.

69. Arnáez, J.; Lana-Renault, N.; Lasanta, T.; Ruiz-Flaño, P.; Castroviejo, J. Effects of farming terraces on hydrological and geomorphological processes. A review. *Catena* **2015**, *128*, 122–134. [CrossRef]

70. Moreno-de-las-Heras, M.; Lindenberger, F.; Latron, J.; Lana-Renault, N.; Llorens, P.; Arnáez, J.; Romero-Díaz, A.; Gallart, F. Hydro-geomorphological consequences of the abandonment of agricultural terraces in the mediterranean region: Key controlling factors and landscape stability patterns. *Geomorphology* **2019**, *333*, 73–91. [CrossRef]

71. Gallart, F.; Llorens, P.; Latron, J. Studying the role of old agricultural terraces on runoff generation in a small mediterranean mountainous basin. *J. Hydrol.* **1994**, *159*, 291–303. [CrossRef]

72. Meerkerk, A.L.; van Wesemael, B.; Bellin, N. Application of connectivity theory to model the impact of terrace failure on runoff in semi-arid catchments. *Hydrol. Process.* **2009**, *23*, 2792–2803. [CrossRef]

73. Calsamiglia, A.; Fortesa, J.; García-Comendador, J.; Lucas-Borja, M.E.; Calvo-Cases, A.; Estrany, J. Spatial patterns of sediment connectivity in terraced lands: Anthropogenic controls of catchment sensitivity. *Land Degrad. Dev.* **2018**, *29*, 1198–1210. [CrossRef]

74. Brandolini, P.; Cevasco, A.; Capolongo, D.; Pepe, G.; Lovergine, F.; Del Monte, M. Response of terraced slopes to a very intense rainfall event and relationships with land abandonment: A case study from Cinque Terre (Italy). *Land Degrad. Dev.* **2018**, *29*, 630–642. [CrossRef]

75. Romero-Díaz, A.; Ruiz-Sinoga, J.D.; Robledano-Aymerich, F.; Brevik, E.C.; Cerdà, A. Ecosystem responses to land abandonment in western mediterranean mountains. *Catena* **2017**, *149*, 824–835. [CrossRef]

76. Cerdà, A.; Ackermann, O.; Terol, E.; Rodrigo-Comino, J. Impact of farmland abandonment on water resources and soil conservation in citrus plantations in eastern Spain. *Water* **2019**, *11*, 824. [CrossRef]

77. Seeger, M.; Rodrigo-Comino, J.; Iserloh, T.; Brings, C.; Ries, J.B. Dynamics of runoffand soil erosion on abandoned steep vineyards in the Mosel area, Germany. *Water* **2019**, *11*, 2596. [CrossRef]

78. Pepe, G.; Mandarino, A.; Raso, E.; Scarpellini, P.; Brandolini, P.; Cevasco, A. Investigation on farmland abandonment of terraced slopes using multitemporal data sources comparison and its implication on hydro-geomorphological processes. *Water* **2019**, *11*, 1552. [CrossRef]

79. Lasanta, T.; Arnaéz, J.; Flaño, P.R.; Monreal, N.L.-R. Agricultural terraces in the Spanish mountains: An abandoned landscape and a potential resource. Los bancales en las montañas españolas: Un paisaje abandonado y un recurso potencial. *Bol. La Asoc. Geogr. Esp.* **2013**, *63*, 301–322.

80. Zoumides, C.; Bruggeman, A.; Giannakis, E.; Camera, C.; Djuma, H.; Eliades, M.; Charalambous, K. Community-based rehabilitation of mountain terraces in Cyprus. *Land Degrad. Dev.* **2017**, *28*, 95–105. [CrossRef]

81. Nadal-Romero, E.; Khorchani, M.; Lasanta, T.; García-Ruiz, J.M. Runoff and solute outputs under Different land mountain experimental station. *Water* **2019**, *11*, 976. [CrossRef]

82. Lucas-Borja, M.E.; Zema, D.A.; Antonio Plaza-Álvarez, P.; Zupanc, V.; Baartman, J.; Sagra, J.; González-Romero, J.; Moya, D.; de las Heras, J. Effects of different land uses (abandoned farmland, intensive agriculture and forest) on soil hydrological properties in southern Spain. *Water* **2019**, *11*, 503. [CrossRef]

83. Solomou, A.D.; Proutsos, N.D.; Karetsos, G.; Tsagari, K. Effects of climate change on vegetation in mediterranean forests: A review. *Int. J. Environ. Agric. Biotechnol.* **2017**, *2*, 240–247. [CrossRef]

84. Kelly, E.F.; Chadwick, O.A.; Helinski, T.E. The effect of plants on mineral weathering. *Biogeochemistry* **1998**, *42*, 21–53. [CrossRef]

85. Andréassian, V. Waters and forests: From historical controversy to scientific debate. *J. Hydrol.* **2004**, *291*, 1–27. [CrossRef]

86. Ortigosa, L.M.; Garcia-Ruiz, J.M.; Gil-Pelegrin, E. Land reclamation by reforestation in the central Pyrenees. *Mt. Res. Dev.* **1990**, *10*, 281–288. [CrossRef]

87. Nadal-Romero, E.; Cammeraat, E.; Serrano-Muela, M.P.; Lana-Renault, N.; Regüés, D. Hydrological response of an afforested catchment in a mediterranean humid mountain area: A comparative study with a natural forest. *Hydrol. Process.* **2016**, *30*, 2717–2733. [CrossRef]

88. García-Ruiz, J.M.; Lasanta, T.; Nadal-Romero, E.; Lana-Renault, N.; Álvarez-Farizo, B. Rewilding vs. restoring. cultural landscapes in mediterranean mountains: Opportunities and challenges. *Land Use Policy* **2020**, under revision.

89. Nogués-Bravo, D.; Simberloff, D.; Rahbek, C.; Sanders, N.J. Rewilding is the new Pandora's box in conservation. *Curr. Biol.* **2016**, *26*, 87–91. [CrossRef] [PubMed]

90. Nabuurs, G.-J.; Schelhaas, M.-J.; Mohren, G.; Frits, M.J.; Field, C.B. Temporal evolution of the European forest sector carbon sink from 1950 to 1999. *Glob. Chang. Biol.* **2003**, *9*, 152–160. [CrossRef]

91. Nadal-Romero, E.; Cammeraat, E.; Pérez-Cardiel, E.; Lasanta, T. How do soil organic carbon stocks change after cropland abandonment in mediterranean humid mountain areas? *Sci. Total Environ.* **2016**, *566*, 741–752. [CrossRef]

92. Soulé, M.; Noss, R. Ewilding and biodiversity: Complementary goals for continental conservation. *Wild Earth* **1998**, *8*, 19–29.

93. Queiroz, C.; Beilin, R.; Folke, C.; Lindborg, R. Farmland abandonment: Threat or opportunity for biodiversity conservation? A global review. *Front. Ecol. Environ.* **2014**, *12*, 288–296. [CrossRef]

94. Navarro, L.M.; Pereira, H.M. Rewilding abandoned landscapes in Europe. *Ecosystems* **2012**, *15*, 900–912. [CrossRef]

95. Tarolli, P.; Preti, F.; Romano, N. Terraced landscapes: From an old best practice to a potential hazard for soil degradation due to land abandonment. *Anthropocene* **2014**, *6*, 10–25. [CrossRef]

96. Lasanta, T.; Khorchani, M.; Pérez-Cabello, F.; Errea, P.; Sáenz-Blanco, R.; Nadal-Romero, E. Clearing shrubland and extensive livestock farming: Active prevention to control wildfires in the mediterranean mountains. *J. Environ. Manag.* **2018**, *227*, 256–266. [CrossRef] [PubMed]

 *water* 

Article

Effects of Different Land Uses (Abandoned Farmland, Intensive Agriculture and Forest) on Soil Hydrological Properties in Southern Spain

Manuel Esteban Lucas-Borja [1], Demetrio Antonio Zema [2,*], Pedro Antonio Plaza-Álvarez [1], Vesna Zupanc [3], Jantiene Baartman [4], Javier Sagra [1], Javier González-Romero [1], Daniel Moya [1] and Jorge de las Heras [1]

[1] Escuela Técnica Superior Ingenieros Agrónomos y Montes, Universidad de Castilla-La Mancha, Campus Universitario, E-02071 Albacete, Spain; manuelestaban.lucas@uclm.es (M.E.L.-B.); pedro.plaza@uclm.es (P.A.P.-Á.); javier.sagra@uclm.es (J.S.); Javier.Gonzalez@uclm.es (J.G.-R.); Daniel.Moya@uclm.es (D.M.); Jorge.Heras@uclm.es (J.d.l.H.)

[2] Department AGRARIA, Mediterranean University of Reggio Calabria, Loc. Feo di Vito, I-89122 Reggio Calabria, Italy

[3] Department of Agronomy, Biotechnical Faculty, University of Ljubljana, Jamnikarjeva 101, 1000 Ljubljana, Slovenia; vesna.zupanc@bf.uni-lj.si

[4] Soil Physics and Land Management Group, Wageningen University, 6708 WG Wageningen, The Netherlands; jantiene.baartman@wur.nl

* Correspondence: dzema@unirc.it; Tel.: +39-0965-1694295

Received: 2 February 2019; Accepted: 6 March 2019; Published: 11 March 2019

Abstract: A detailed knowledge of soil water repellency (SWR) and water infiltration capacity of soils under different land uses is of fundamental importance in Mediterranean areas, since these areas are prone to soil degradation risks (e.g., erosion, runoff of polluting compounds) as a response to different hydrological processes. The present study evaluates the effects of land uses on SWR and soil hydraulic conductivity (SHC) by direct measurements at the plot scale in three areas representing (1) intensive agricultural use, (2) abandoned farmland, and (3) a forest ecosystem in Southern Spain under Mediterranean climatic conditions. The physico-chemical properties and water content of the experimental soils were also measured. Significant SWR and SHC differences were found among the analyzed land uses. Forest soils showed high SWR and low SHC, while the reverse effects (that is, low SWR and high SHC) were detected in soils subjected to intensive agriculture. Organic matter and bulk density were important soil properties influencing SWR and SHC. The study, demonstrating how land uses can have important effects on the hydrological characteristics of soils, give land managers insights into the choice of the most suitable land use planning strategies in view of facing the high runoff and erosion rates typical of the Mediterranean areas.

Keywords: soil water repellency; soil hydrological conductivity; soil physico-chemical properties; vegetal cover; vegetation cover

---

1. Introduction

Mediterranean areas are very prone to soil degradation risks (e.g., surface runoff, erosion, transport of nutrients and other polluting compounds): the soils are generally shallow with low levels of organic matter, low aggregate stability, and nutrient content [1], and the climate is characterized by frequent and intense rainstorms producing a high magnitude of flash floods with high erosive power [2]. This combination of soil type and climate leads to a peculiar hydrological response with high runoff rates that have high erosive power. This response is also affected by both land use and soil cover and their temporal and spatial variability, which are considered the most important factors affecting the intensity and frequency of surface runoff and soil erosion [3,4]. Inappropriate land use or a poor soil cover may accelerate water runoff and soil erosion dynamics [4,5], leading to unsustainable land degradation processes. The main causes of such negative environmental impacts are agricultural practices, deforestation, overgrazing, land abandonment, wildfires, and civil works [6,7]. For instance, agriculture is thought to generate high erosion rates [4,8,9], and the use of land for intensive agriculture, in general, may cause soil damage [10,11]. As well known, agriculture, particularly when intensive and subject to frequent tillage operations, may strongly affect the physico-chemical properties of soil, making it more prone to erosion and quality decay.

Erosion problems have also been associated with land abandonment of both agricultural and marginal farmlands [1,12] with particular reference to the Mediterranean of Western Europe [9,13,14]. For instance, in SE Spain, where drastic human impacts were recorded in the second half of the past century [15], anthropogenic land use changes have triggered soil erosion and led to severe land degradation [1]. Here, since the mid-twentieth century, the use of land for intensive agriculture and the gradual process of vegetation recovery of abandoned farmlands and marginal areas have been the main causes of land degradation [10,11]. These land use changes, whose environmental impacts are worsened by the specific climate and the soil fragility, make this region very prone to runoff generation and soil erosion [3] with consequent pollution of water bodies, soil organic matter decline, devastating floods, reservoir siltation, and mass failures [1,15].

In Mediterranean areas, the hydrological processes generating water runoff, soil erosion, and transport of polluting compounds are dominated by the infiltration-excess mechanism [16]. In soils of the semi-arid Mediterranean climate, exhibiting low hydraulic conductivity, surface runoff, and soil erosion, can be high. Moreover, such soils could be expected to also be affected by water repellency [17,18], further decreasing infiltration rates [19], which in turn leads to increased runoff and erosion [20,21], to accelerated leaching of agrochemicals [22], and to a reduction in the vegetal cover of soils, leaving the latter bare and thus prone to erosion [23,24]. Thus, depending on the water repellency level, Mediterranean soils may have an infiltration rate of up to several orders of magnitude lower than would be expected (e.g., [20,21,25,26]). It is thus evident how a detailed knowledge of the soil water repellency (hereinafter SWR) and water infiltration capacity (in terms of soil hydraulic conductivity, hereinafter SHC) under different land uses is of fundamental importance in Mediterranean areas to control hydrological risks and other environmental impacts linked to these risks.

SWR has been studied worldwide [20,21], in both forest [27] and agricultural soils [24,28]. The agricultural soils are usually considered wettable [29] and thus a little subject to SWR, while other studies have demonstrated that some management practices may induce SWR in cultivated soils [30–32]. The attention paid to surface runoff and soil erosion rates in marginal and abandoned lands (e.g., [33]) has highlighted sometimes contrasting results in the Mediterranean [34] and, consequently, the difficulty to fully understand the effects of land abandonment on hydrological response and soil erosion [9], as modified by both SWR and SHC. Therefore, it is important to know about repellency and infiltration under typical land uses in the drier parts of the Mediterranean basin [18]. A better comprehension of these fundamental soil parameters is important for both agricultural production and protection of abandoned farmland from hydro-geological risks [24].

This study evaluates the effects of land use on SWR and SHC by direct measurements at the plot scale in three areas representing (1) intensive agricultural use, (2) abandoned farmland, and (3) a forest ecosystem in Southern Spain. In addition, the physico-chemical properties and water content, which, as it is well known, can influence SWR and SHC, were also measured on these representative Mediterranean soils. The objective of the study is to evaluate which of the analyzed land uses (agriculture, abandoned farmland and forest) shows the highest SWR and lowest SHC in the experimental conditions, also linking these properties to important soil characteristics. By demonstrating how land uses can have important effects on the hydrological characteristics of soils, we want to give land managers insights on the choice of the most suitable land use planning in view of facing the high runoff and erosion rates typical of the Mediterranean areas.

2. Materials and Methods

2.1. Study Area

The study area is located in the municipality of Villamalea (39.36422° N, −1.59689° E, Albacete, SE Spain) (Figure 1). The climate is typically Mediterranean, "*Csa*", according to the Köppen-Geiger classification [35]. The average annual rainfall and temperature are 407 mm and 14 °C, respectively (Figure 2). The rainfall is mainly distributed in spring and autumn, with a long drought in summer that usually lasts from June–September.

Elevation of the studied plots ranges between 760 and 770 m a.s.l., a flat terrain, typical for the Iberian Plateau (Meseta). The plateau is characterized by high elevated (>600 m a.s.l.) region with an undulating (hilly) landscape that is prone to high erosion rates due to the dry climate conditions that reduce the vegetation cover.

The study area is a traditional Mediterranean forest and agricultural land, where pine plantations, natural forest, almonds, olive, cereals, and vineyards are widespread (Figure 1). The natural tree vegetation of the study area mainly consists of *Pinus pinea* L. and *Pinus halepensis* M. The main shrubs and herbaceous species found at the study site are *Rosmarinus officinalis* L., *Brachypodium retusum* (Pers.) Beauv., *Lavandula latifolia* Medik., *Thymus vulgaris* L., *Stipa tenacissima* (L.), *Quercus coccifera* L. and *Plantago albicans* L. Agriculture consists mainly of vineyards, olive orchards and maize crops, with vine production providing the main agricultural income in the study region.

Figure 1. Location of the study area (**A**, Villamalea, Castilla La Mancha, SE Spain) with the typical land uses (**B**: Forest; **C**: Intensive agriculture; **D**: Abandoned farmland).

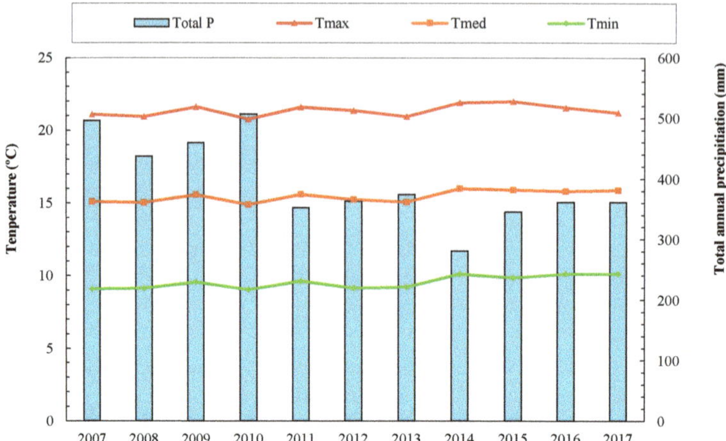

Figure 2. Annual precipitation (total P, mm) as well as mean maximum, average and minimum air temperatures (T_{max}, T_{med} and T_{min}) in the study area (Villamalea, Castilla La Mancha, SE Spain).

According to FAO-UNESCO [36] and the IUSS Working Group WRB [37], the soils of the study area are classified as *Calcic cambisols*. Organic matter content is on average 1.1 g/m^3, varying from 1.05–1.12 g/m^3. The texture is sandy clay loam (60% sand, 10% silt, and 27% clay) over a limestone parent material.

2.2. Experimental Design

2.2.1. Plot Description

In autumn 2018, nine 10 × 10 m^2 plots were set up in each of the three land uses (27 plots in total) of the same study area with the typical climate and soil characteristics described in Section 2.1. The analyzed land uses were (1) intensive agriculture; (2) abandoned farmland; (3) natural forest. The plots with similar slopes (1–5%) were distributed selecting certain site characteristics, slopes, and aspects to ensure comparability among the 27 plots. Distances between plots were always over 500 m.

The land use related to intensive agriculture (hereinafter "IA") consisted of an olive grove about 20 years old. Cropping operations follow the usual standards of the local farmers. The orchard is tilled three times per year, scheduled depending on the soil moisture and weed removal needs. Tillage is carried out just before or on the occasion of weed germination in order to keep the soil surface clean and tidy and is refined by eliminating herbs near the trunk of each tree (in June or July) by hoes. No fertilizers are used in the olive grove. Residues of pruning, operated in February or March, are concentrated and burned at the margins of the grove, but not chipped, as usually made for other crops (e.g., citrus and apricots).

The abandoned farmland (hereinafter "AF") was an olive orchard, abandoned about 15 years ago and now mainly covered by herbs and shrubs (mainly *Rosmarinus officinalis* L., *Brachypodium retusum* (Pers.) Beauv., *Lavandula latifolia* Medik., *Thymus vulgaris* L., *Macrochloa tenacissima* L., *Quercus coccifera* L., *Plantago albicans* L. *Eryngium campestre* L., and *Pistacia lentiscus* L.).

Forest plots (hereinafter "FO") are covered by *Pinus halepenis* M., about 50 years old, with a density of 560 trees per ha. Shrubs found underneath the trees are *Quercus coccifera* L. and *Macrochloa tenacissima* L.

2.2.2. Soil Property Measurements

Herbal cover, rock fragments, dead woody matter, and bare soil covers were measured at three 5-m transects in each plot, measuring the percent cover of each property over a grid of 1 m × 1 m.

Bulk density was calculated on triple samples per plot as the weight of soil in a given volume of the core extracted by a small cylinder at a depth of 20–30 cm. Soil water content (SWC) was estimated using a HOBOnet Soil Moisture Sensor, which integrates the field-proven ECH2O™ EC5 Sensor and provides readings directly in volumetric water content. SWC was measured hourly for 24 h on the occasion of the SWR and SHC surveys and the measures were then averaged.

The main topsoil properties (0–5 cm) were determined on six samples per plot. A total of 162 soil samples (6 samples × 9 plots × 3 land uses) were collected. Soil characterization was carried out by measuring the following parameters: (i) texture (sand, silt and clay percentage) following the methodology of Guitian and Carballas (1976) [38]; (ii) soil organic matter (OM), estimated from organic carbon following the methodology proposed by Nelson and Sommers (1996) [39]; (iii) electrical conductivity (EC) and pH, measured in deionized water (1:2.5 and 1:5 w:w, correspondingly) at 20 °C [40]; (iv) content of (total nitrogen, using the Kjeldahl method [41]) and available phosphorus [42]); (v) content of potassium; magnesium; sodium; and calcium, exchangeable cations measured using the barium-chloridetriethanolamine method [43]).

Based on the parameters above, the carbon-to-nitrogen ratio (C/N) and cation exchange capacity (CEC) were calculated.

The water infiltration capacity of soils was estimated measuring SHC by a MiniDisk infiltrometer (MDI). In more detail, first the measured cumulative infiltration values (I, [m]) were fitted against the measurement intervals (t, [s]), both given by MDI, using Equation (1):

$$I = C_1 t + C_2 \sqrt{t} \tag{1}$$

and the coefficients C_1 [m/s] and C_2 [m/s$^{1/2}$] were estimated by interpolation. Coefficient C_1 is related to SHC, and C_2 is the sorptivity [44]. Then, SHC (k, [mm/h]) was calculated using the following equation:

$$k = \frac{C_1}{A} \tag{2}$$

where coefficient A is a value relating the Van Genuchten parameters (n and α) for a certain soil type to the suction rate (h_0) and the infiltrometer disk radius (2.25 cm). Entering the values of n, α, and h_0 (assumed in this study to be equal to −2 cm) of the experimental soils in the table reported in the MDI manual [44], a value of 2.8 was derived for A. Equations (1) and (2) were proposed by Reference [45].

SWR was measured as follows: 15 drops of distilled water were released, using a pipette, on the soil surface of a 1-m transect, to homogenize the changing soil conditions; the time necessary for drops to infiltrate completely into the soil was measured by a stopwatch. Before measurement, litter cover was removed, and the soil surface was cleaned using a brush. The high number of replicates (ten points per plot) assured the best reliability for this measurement. The method used is recognized as one of the most appropriate for evaluating the SWR degree in field measurements [46].

2.3. Statistical Analysis

A one-way ANOVA (using land use as the independent factor) was applied to evaluate the statistical significance of the variations in soil cover, physico-chemical properties, SWR and SHC. All the plots were considered spatially independent. An independent Fisher's minimum significant difference test (LSD) was used for the post hoc analysis comparisons. A $p < 0.05$ level of significance was adopted. It was not necessary to perform data transformations for the analysis. ANOVA assumes normality and this assumption was checked using QQ-plots. All measured variables were used to perform the principal component analysis (PCA), the latter being based on a Spearman rank correlation matrix, to reduce the dimensionality of the data set. The statistical analysis was performed by version 3.24 of the R Project for Statistical Computing.

3. Results

The soil physical characteristics as measured for the three land uses are shown in Table 1. The texture of the experimental soils was statistically similar between AF and IA; the texture of the FO soil was significantly different from the other land uses. In general, the soil was finer in the AF and IA land uses (clay loam and loam, respectively) and coarser in FO. Soils subject to IA were more compacted (mean bulk density of 1.25 g cm^{-3}), while the FO soils had a lower bulk density (0.83 g cm^{-3}) (Table 1).

Table 1. Soil physical characteristics (mean ± standard deviation) in the study area (Villamalea, Castilla La Mancha, SE Spain). Different lower-case letters indicate statistically significant differences following LSD test (*P*-value < 0.05).

Land Use *	Sand (%)	Silt (%)	Clay (%)	Bulk Density (g cm^{-3})	Rock (%)	Herbal Cover (%)	Bare Soil (%)	Dead Woody Matter (%)
IA	46.0 ± 9.9 [b]	35.3 ± 8.7 [a]	18.7 ± 9.7 [ab]	1.25 ± 0.2 [a]	14.2 ± 2.9 [a]	1.0 ± 0.7 [b]	72.8 ± 3.8 [a]	33.3 ± 5.7 [b]
AF	32.9 ± 12.2 [b]	39.8 ± 7.9 [a]	27.7 ± 6.8 [a]	1.10 ± 0.3 [b]	3.1 ± 2.2 [b]	32.4 ± 4.9 [a]	31.1 ± 2.8 [b]	12.8 ± 5.7 [c]
FO	70.0 ± 10.1 [a]	19.5 ± 9.5 [b]	10.3 ± 6.7 [b]	0.83 ± 0.2 [c]	10.0 ± 2.2 [ab]	29.5 ± 5.5 [a]	1.3 ± 0.5 [c]	59.1 ± 6.6 [a]

* AF: Abandoned Farmland; IA: Intensive Agriculture; FO: Forest.

The IA showed the lowest vegetation cover (on average only 1% of the total sampled area). This latter parameter was similar between the other land uses (about 30% and not significantly different between abandoned farmland and FO). On average, only 1% of the soil in the FO plots was bare against more than 70% in IA (Table 1). The percentage of dead woody matter (plant material less than 2 mm in diameter) was statistically significantly different for all land uses, being highest in FO plots, followed by IA and AF soils.

The soil chemical characteristics as measured in the three land uses are shown in Table 2. The values of pH (on average 8.5–8.7) and EC (0.21–0.24 mS cm^{-1}) were not significantly different among the investigated land uses. The OM content, in tune with the soil total carbon, was similar between AF and FO (mean values between 1.3 and 1.5%), and significantly lower in IA (on average 0.6%). OM was not correlated to dead woody matter, presumably due to the different mineralization levels of the investigated soils, on which the previous and current tillage practices may have played a role. Also, the contents of P, K, and cations (Na$^+$, Ca^{++} and Mg^{++}) were significantly different between land uses. FO soils had the lowest concentrations (except for Ca^{++}) and the AF the highest (except for P). Conversely, there were no statistically significant differences in N content (on average 0.05–0.1%). Due to this, the C/N ratio was quite similar among all the studied land uses (mean values between 11.9 and 16.3), in spite of the differences recorded in C mean contents (from 0.6% of IA to 1.5% of AF). Finally, CEC was significantly higher in AF (on average 16.3 meq/100 g of soil), while FO soils showed the lowest value (6.8 meq/100 g, this latter not significantly different from IA, 9.9 meq/100 g) (Table 2).

Table 2. Soil chemical characteristics (mean ± standard deviation) in the study area (Villamalea, Castilla La Mancha, SE Spain). Different lower-case letters indicate statistically significant differences following the LSD test (*P*-value < 0.05).

Land Use *	pH (-)	OM (%)	C (%)	N (%)	P (ppm)	K (meq/100 g)	Na (meq/100 g)
IA	8.7 ± 0.7 [a]	0.98 ± 1.0 [b]	0.56 ± 0.2 [b]	0.05 ± 0.01 [a]	21.7 ± 6.3 [a]	0.5 ± 0.1 [b]	0.32 ± 0.11 [a]
AF	8.5 ± 0.7 [a]	2.7 ± 0.9 [a]	1.5 ± 0.5 [a]	0.10 ± 0.05 [a]	8.1 ± 2.1 [b]	0.8 ± 0.2 [a]	0.35 ± 0.10 [a]
FO	8.7 ± 0.6 [a]	2.2 ± 0.8 [a]	1.3 ± 0.5 [a]	0.08 ± 0.03 [a]	7.4 ± 1.1 [b]	0.2 ± 0.1 [c]	0.10 ± 0.05 [b]

Land Use *	Ca (meq/100 g)	Mg (meq/100 g)	C/N	K/Mg	Ca/Mg	CEC (meq/100 g)	EC (mS/cm)
IA	32.1 ± 5.9 [b]	1.3 ± 0.3 [a]	11.9 ± 3.7 [a]	0.3 ± 0.1 [a]	24.1 ± 5.9 [b]	9.9 ± 3.8 [ab]	0.21 ± 0.03 [a]
AF	41.5 ± 7.9 [a]	1.6 ± 0.3 [a]	14.2 ± 3.5 [a]	0.5 ± 0.1 [a]	25.8 ± 6.2 [b]	16.3 ± 5.3 [a]	0.21 ± 0.02 [a]
FO	32.2 ± 9.4 [a]	0.7 ± 0.2 [b]	16.3 ± 8.2 [a]	0.3 ± 0.1 [a]	43.9 ± 7.7 [a]	6.8 ± 3.8 [b]	0.24 ± 0.04 [a]

* AF: Abandoned Farmland; IA: Intensive Agriculture; FO: Forest; OM: Organic Matter; CEC: Cation Exchange Capacity; EC: Electrical Conductivity; C: Total carbon; N: Total nitrogen.

The analysis of the hydrological properties of the investigated soils showed much higher mean SWR values on soil surface compared to measurements made at 2 cm below ground (4-fold, AF, and 20-fold, FO). Only in soils under IA, the surface, and below ground SWR were similar. Moreover, FO soils had the highest SWR at both investigated depth (on average 74 s, on the soil surface and 3.5 s, 2 cm below the ground), while the lowest values were measured in IA (2.5 s) on soil surface, and AF (2.0 s) at 2 cm below the ground. Soil water content was 10.0% for AF and FO and 15.2% for IA) at the date of measurements (Figure 3a).

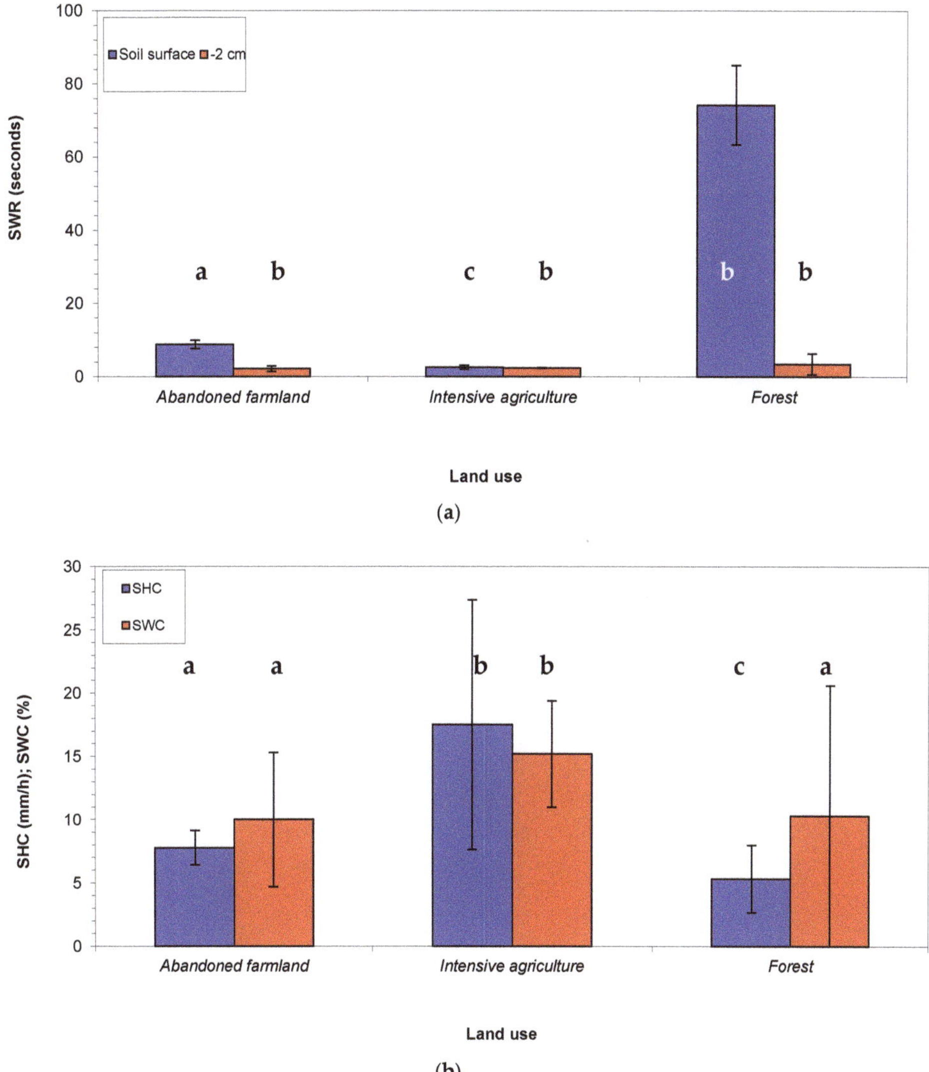

Figure 3. Soil water repellency (SWR, measured on the soil surface and at 2 cm below ground, (**a**), water content (SWC, **b**) and hydraulic conductivity (SHC, **b**) in the study area (Villamalea, Castilla La Mancha, SE Spain). Different lower-case letters indicate statistically significant differences comparing land uses for SWC and SHC following LSD test (*P*-value < 0.05).

In soils under IA, the highest SHC was detected (on average 17 mm/h), while the FO soils showed the lowest values (5 mm/h), similar (but significantly different) from SHC measured in the AF (mean value of 7 mm/h) (Figure 3b and Table 3).

Table 3. Results of ANOVA applied to land uses (Abandoned Farmland; Intensive Agriculture; Forest) for soil water content (SWC), repellency (SWR) and hydraulic conductivity (SHC) in the study area (Villamalea, Castilla La Mancha, SE Spain).

Soil Parameter *	Degree of Freedoms	F-Ratio	P-Value
SWC		28.9	
SWR (soil surface)		53.7	
SWR (−2 cm)	2	34.7	<0.05
SHC		10.5	

* SWC = Soil Water Content; SWR = Soil Water Repellency; SHC = Soil Hydraulic Conductivity.

The PCA evidenced a clear clustering of the three investigated land uses with regard to the properties of soils, assumed as original variables (Figure 4). The PC_1 and PC_2 explained 51% and 37% (in total 88%), respectively, of the total variance of the original variables (Table 4). EC, pH, CEC, and nutrient contents were the variables with the higher loading factors on PC_1, whereas plot % of vegetation cover, bare soil, dead woody matter, total carbon content, and bulk density showed the higher loadings on PC_2.

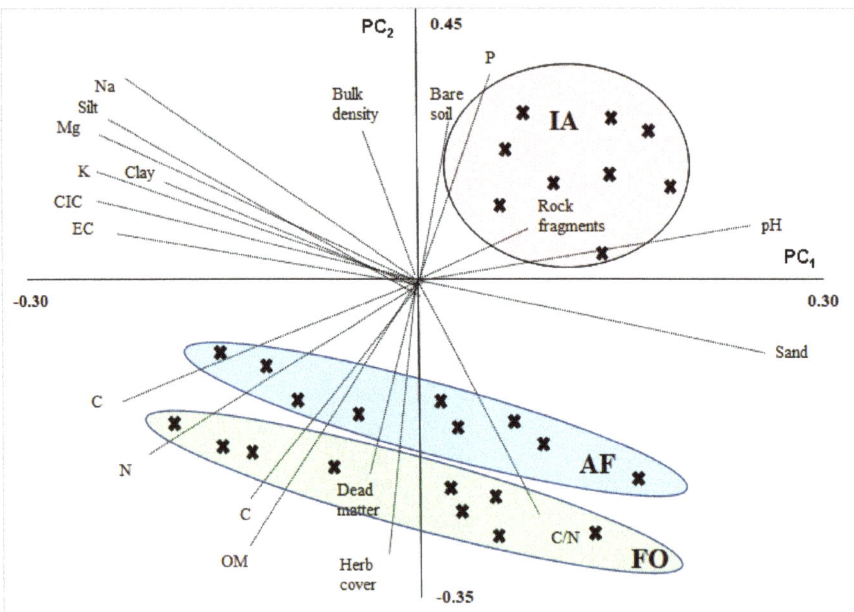

Figure 4. Principal component analysis applied to properties of soils subject to different land uses (AF: Abandoned Farmland; IA: Intensive Agriculture; FO: Forest) in the study area (Villamalea, Castilla La Mancha, SE Spain) (the ellipses represent the evident clusters achieved by coupling ANOVA and PCA and using land uses as factors).

Table 4. Factor loadings on the first three principal components (PC) applied to properties of soils subjected to different land uses in the study area (Villamalea, Castilla La Mancha, SE Spain).

Soil Properties (Original Variables)	PCs	
	PC$_1$ (51%)	PC$_2$ (37%)
Rock	0.111	0.113
Vegetation cover	−0.102	−0.263
Bare soil	0.088	0.347
Dead woodymatter	−0.070	−0.268
Bulk density	0.009	0.306
Sand content	0.269	−0.202
Silt content	−0.244	0.245
Clay content	−0.293	0.141
pH	0.311	0.054
Electrical conductivity	−0.312	0.051
Organic matter	−0.191	−0.304
Total nitrogen content	−0.247	−0.237
Phosphorous content	0.097	0.365
Potassium content	−0.307	0.089
Sodium content	−0.230	0.264
Calcium content	−0.311	−0.059
Magnesium content	−0.263	0.214
Total carbon content	−0.193	−0.303
Cation Exchange Capacity	−0.310	0.070

Note: in parentheses the percentage of the total variance explained by each PC is reported.

4. Discussion

Many studies have demonstrated how and by what extent land uses influence the physico-chemical and hydrological characteristics of soil (e.g., [4,11,16,34,47]). However, these characteristics also depend on the specific soil properties, such as texture and vegetal cover. For the investigated soils (sandy to clayey texture with variable vegetation cover), it has been shown that some of the chemical characteristics (namely pH, EC, N content and the C/N ratio) were very similar among the investigated land uses. Conversely, significant gradients in the OM content and CEC, two very important soil properties regarding soil fertility and productivity, were noticed when comparing AF (showing the highest values) to soils subject to IA. This soil depletion in OM and CEC may be attributable to plant uptake, due to the crop growth [48]. Moreover, soil under IA has lower OM and higher C/N ratio on average, although not optimal [49]; the other two land uses (AF and FO) have much less favorable C/N ratio for plants.

The most important variations among the investigated land uses were noticed in the soil hydrological properties: SWR and SHC varied significantly among the experimental land uses. More specifically, the topsoil of forested land had a much higher SWR compared to the surface soils subject to IA and AF. Even a slight SWR of FO soil may have noticeable effects on water infiltration rates, generating more runoff—particularly in summer, when the soil is drier—and thus enhancing soil erosion [27]. However, although it has been reported that SWR is particularly common under rangeland or forestland such as permanent grassland and deciduous shrub and forest terrain [50–52], it can also occur on agricultural land [27]. SWR affects both coarse and fine-textured soils [21,29,53] and occurs at low to moderate water contents [26,54]. In particular, coarse-textured soils, as those of the forestland analyzed in this study, are more prone to water repellency, even when they are permanently vegetated [55]. The level of SWR depends on the soil particle fraction with a hydrophobic surface coating [52] and is influenced by the surface area of the particles, which varies considerably with soil texture [56]. Therefore, in the sandy soils of our forestland, which have the lowest surface area, a hydrophobic surface impacts a larger proportion of particles than for a loamy or clayey soil where the surface area is up to three orders of magnitude greater [56,57]. Moreover, the plant species surveyed in the forest of this study (eucalyptus and pines) are perennial trees with a considerable concentration of resins, waxes or aromatic oils, which have commonly been associated with SWR [58–60]. Under these conditions, it may be advisable to develop proper strategies to reduce SWR, such as a more effective soil management, the addition of clays to increase particle surface area, tillage to break-up and abrade hydrophobic surfaces and the use of chemical wetting agents [29,61]. Moreover, regions with a Mediterranean climate with prolonged dry periods, such as southern Spain, could be particularly affected by SWR and its hydrological impacts, which bring soils within a water content range in which SWR is exhibited [18,62,63].

Conversely, SWR levels surveyed in the other two investigated land uses (IA and AF) are less severe. In the soils previously subjected to agricultural activities, plant natural succession after land abandonment helps to avoid or reduce SWR, contributing to water penetration into the deeper soil layers, thanks to preferential flow paths via plant roots and stem flow [64]. Regarding the plots subjected to IA, agricultural soils are usually considered wettable [29], even though some studies have found that soil management practices can induce water repellency also in cropped areas [30–32]. In general, tilled sites are virtually unaffected by SWR [55]; furthermore, cultivation promotes rainfall infiltration, and, as a consequence, the runoff and erosion rates are significantly reduced [65]. However, although runoff and erosion are mostly reduced after plowing, these processes may increase again over time because of crusting, especially in silty soils [66].

Beside the low SWR levels, the reduced SHC of forestland detected in this study, compared to soils subject to IA, deserves much attention. As a matter of fact, depending on the severity of SWR, such soils may have water infiltration rates lower than would be expected on the basis of their pore size distribution and SWC (e.g., [20,25]). The combination of a reduced wettability (due to high SWR) and low SHC may enhance surface runoff (e.g., [26,67]). With this type of land use, the exposure of soils to the highest and most erosive rainfalls in the experiment area during autumn/winter may aggravate the erosive risks. However, this particular hydrological response of forest sandy soils can be prevented by a proper ground cover and by natural vegetation, which, having a strong influence on soil hydrological properties, reduces soil erosion rates [68].

Also, in the AF of this study, a lower SHC was detected (of the same order of magnitude as FO), which may lead to the degradation of the ecosystem at least during the first 3–5 years after abandonment [65]. This reduced SHC could be mainly due to the finer texture of AF soil compared to IA, but also the land use may have had a role. However, it has been demonstrated that, after land abandonment, the vegetation recovery should improve soil water-retention capacity and hydraulic conductivity, thus increasing infiltration and decreasing runoff and erosion rates [65,69]. In such a way, plant re-growth starts to control the hydrological and erosional soil response [64]. Moreover, the C/N ratio of AF soils is less favorable than in agricultural soils (and this could be due to the absence of fertilizer applications), which subsequently influenced plant succession (depending on the duration of abandonment) [49].

Overall, the worse hydrological response of FO soils compared to IA and AF plots, detected in this study, suggest that, although the forest tree planting by reforestation has been adopted as a viable land management strategy in many parts of the Mediterranean basin, natural scrubland, similar as those of revegetation processes in AF, may be more appropriate with regard to water use efficiency and soil conservation measures, since they assure a lower SWR and a slightly higher SHC [18].

However, it must be understood that the SHC measurements of the tension infiltrometer carried out in this study relate to the unsaturated soil. This parameter may be initially more important than saturated hydraulic conductivity at the time scale of convective rainfall, which is typically short and only lasts 20–60 min, common in many Mediterranean areas [70]. Since this investigation showed that soils of forestland and AF are affected by a lower SHC compared to agricultural areas, these land uses require caution in semi-arid or arid Mediterranean ecosystems, where runoff and soil erosion risks may be high. As a matter of fact, given that in the soils typical of this climatic context (in particular, those of prevalent sandy texture) the Hortonian (that is, infiltration-excess) overland flow type dominates over concentrated runoff, a reduction of SHC may worsen the hydrological behavior of soils subjected to these land uses, with a possible increase of water runoff and soil erosion [71]. In fact, a lower water infiltration rate increase reduces the water storage of soil during heavy storms, thus increasing the share of precipitation, which generates surface runoff [16,47].

5. Conclusions

The effects of land use (abandoned farmland, intensive agriculture and natural forest) have been evaluated at the plot scale with particular reference to the physico-chemical properties, water content and repellency, and the hydraulic conductivity of soils. While most soil properties were not significantly different between the land uses, the hydraulic properties of the investigated soils showed specific responses to the different land uses or plant covers. Forest soils showed -high water repellency and low infiltrability, which worsens their hydrological behavior under heavy and frequent storms typical of the Mediterranean landscape. This behavior may increase the risks of soil erosion and pollutant runoff downstream in sloping areas. Abandoned soils previously subjected to agriculture showed a moderate water repellency, but their low hydraulic conductivity can cause serious problems in terms of runoff and soil erosion. However, shrub vegetation recovery, resulting from plant succession, can decrease this concern by increasing soil cover, which may reduce its erodibility. Compared to the forestland and the abandoned land, the agricultural soils were less affected by water repellency and low infiltrability, presumably due to the periodical tillage operations.

Overall, the main conclusion of this study is the important effect of land use on the hydrological characteristics of soils, and indirectly on their different susceptibility to surface runoff and erosion. This suggests paying attention to the specific land use and soil type under the Mediterranean climate (namely, steep sandy soils of forest ecosystems and also in the context of the expected climate changes), which may be affected by high runoff and erosion rates.

Author Contributions: Conceptualization, M.E.L.-B.; methodology, M.E.L.-B.; formal analysis, M.E.L.-B.; investigation, P.A.P.-Á., J.S., J.G.-R., and D.M.; resources, J.d.l.H.; data curation, D.A.Z. and M.E.L.-B.; writing—original draft preparation, D.A.Z.; writing—review and editing, D.A.Z. and M.E.L.-B.; supervision, V.Z. and J.B.; project administration, J.d.l.H.; funding acquisition, J.d.l.H.

Funding: This research received no external funding.

Acknowledgments: This study was supported by funds provided by the University of Castilla-La Mancha to the Forest Ecology Research Group. Also, by the Research Project from the Spanish National Institute for Agricultural and Food Research and Technology (INIA), VIS4FIRE (RTA2017-00042-C05). The authors thanks to Carlos Navarro, Beatriz Ariño and Jose Luis Martínez for the field assistance.

Conflicts of Interest: The authors declare no conflict of interest.

References

1. Cantón, Y.; Solé-Benet, A.; De Vente, J.; Boix-Fayos, C.; Calvo-Cases, A.; Asensio, C.; Puigdefábregas, J. A review of runoff generation and soil erosion across scales in semiarid south-eastern Spain. *J. Arid Environ.* **2011**, *75*, 1254–1261. [CrossRef]

2. Fortugno, D.; Boix-Fayos, C.; Bombino, G.; Denisi, P.; Quiñonero Rubio, J.M.; Tamburino, V.; Zema, D.A. Adjustments in channel morphology due to land-use changes and check dam installation in mountain torrents of Calabria (southern Italy). *Earth Surface Process. Landf.* **2017**, *42*, 2469–2483. [CrossRef]

3. García-Ruiz, J.M. The effects of land uses on soil erosion in Spain: A review. *Catena* **2010**, *81*, 1–11. [CrossRef]

4. Nunes, A.N.; De Almeida, A.C.; Coelho, C.O. Impacts of land use and cover type on runoff and soil erosion in a marginal area of Portugal. *Appl. Geogr.* **2011**, *31*, 687–699. [CrossRef]

5. Dunjó, G.; Pardini, G.; Gispert, M. The role of land use-land cover on runoff generation and sediment yield at a microplot scale, in a small Mediterranean catchment. *J. Arid Environ.* **2004**, *57*, 239–256. [CrossRef]

6. Grimm, M.; Jones, R.; Montanarella, L. *Soil Erosion Risk in Europe*; European Commission; Institute for Environment and Sustainability; European Soil Bureau: Hannover, Germany, 2011.

7.	Yassoglou, N.; Montanarella, L.; Govers, G.; Van Lynden, G.; Jones, R.J.A.; Zdruli, P. *Soil Erosion in Europe. Technical Report for DG XI*; European Soil Bureau: Hannover, Germany, 1998.

8.	Cheng, Q. Soil erosion and ecological reestablishment in the Nianchu river valley in Tibet. *Chin. J. Ecol.* **2002**, *21*, 74–77.

9.	Nunes, A.N.; Coelho, C.O.A.; De Almeida, A.C.; Figueiredo, A. Soil erosion and hydrological response to land abandonment in a central inland area of Portugal. *Land Degrad. Dev.* **2010**, *21*, 260–273. [CrossRef]

10.	Trasar-Cepeda, C.; Leirós, M.C.; Gil-Sotres, F. Hydrolytic enzyme activities in agricultural and forest soils. Some implications for their use as indicators of soil quality. *Soil Biol. Biochem.* **2008**, *40*, 2146–2155. [CrossRef]

11.	Lucas-Borja, M.E.; Calsamiglia, A.; Fortesa, J.; García-Comendador, J.; Guardiola, E.L.; García-Orenes, F.; Gago, J.; Estrany, J. The role of wildfire on soil quality in abandoned terraces of three Mediterranean micro-catchments. *CATENA* **2018**, *170*, 246–256. [CrossRef]

12.	Solé-Benet, A. Spain. In *Soil Erosion in Europe*; Boardman, J., Poesen, J., Eds.; John Wiley & Sons: Chichester, UK, 2006; pp. 311–346.

13.	Lasanta, T.; Garcìa-Ruiz, J.M.; Pérez-Rontomé, C.; Sancho-Marcén, C. Runoff and sediment yield in a semi-arid environment: The effect of the land management after farmland abandonment. *CATENA* **2000**, *38*, 265–278. [CrossRef]

14.	MacDonald, D.; Crabtree, J.R.; Wiesinger, G.; Dax, T.; Stamou, N.; Fleury, P.; Lazpita, J.G.; Gibon, A. Agricultural abandonment in mountain areas of Europe: Environmental consequences and policy response. *J. Environ. Manag.* **2000**, *59*, 47–69. [CrossRef]

15.	Burke, S.M.; Thornes, J.B. A thematic review of EU Mediterranean desertification research in Frameworks III and IV: Preface. *Adv. Environ. Monit. Model* **2004**, *1*, 1–14.

16.	Lucas-Borja, M.E.; Zema, D.A.; Carrà, B.G.; Cerdà, A.; Plaza-Alvarez, P.A.; Cózar, J.S.; Gonzalez-Romero, J.; Moya, D.; de las Heras, J. Short-term changes in infiltration between straw mulched and non-mulched soils after wildfire in Mediterranean forest ecosystems. *Ecol. Eng.* **2018**, *122*, 27–31. [CrossRef]

17.	Trabaud, L. Man and fire: Impacts on Mediterranean vegetation. In *Mediterranean-Type Shrublands*; di Castri, F., Goodall, D.W., Specht, R.L., Eds.; Elsevier: Amsterdam, The Netherlands, 1996; pp. 523–537.

18.	Cerdà, A.; Doerr, S.H. Soil wettability, runoff and erodibility of major dry-Mediterranean land use types on calcareous soils. *Hydrol. Process.* **2007**, *21*, 2325–2336. [CrossRef]

19.	Markus, F.; Hannes, F.; William, A.J.; Leuenberger, J. Susceptibility of soils to preferential flow of water: A field study. *Water Resour. Res.* **1994**, *30*, 1945–1954.

20.	Doerr, S.H.; Shakesby, R.A.; Walsh, R.P.D. Soil hydrophobicity in north-west Europe: Its occurrence and implications for modelling soil hydrological response. In *Second Inter-Celtic Colloquium on Hydrology and the Management of Water Resources, 3–7 July*; British Hydrological Society Occasional Paper; British Hydrological Society: London, UK, 2000; Volume 11, pp. 211–218.

21.	Doerr, S.H.; Shakesby, R.A.; Walsh, R.P.D. Soil water repellency: Its causes, characteristics and hydro-geomorphological significance. *Earth-Sci. Rev.* **2000**, *51*, 33–65. [CrossRef]

22.	Taumer, K.; Stoffregen, H.; Wessolek, G. Seasonal dynamics of preferential flow in a water repellent soil. *Vadose Zone, J.* **2006**, *5*, 405–411. [CrossRef]

23.	McKissock, I.; Gilkes, R.J.; Harper, R.J.; Carter, D.J. Relationships of water repellency to soil properties for different spatial scales of study. *Aust. J. Soil Res.* **1998**, *36*, 495–507. [CrossRef]

24.	Brevik, E.C.; Cerdà, A.; Mataix-Solera, J.; Pereg, L.; Quinton, J.N.; Six, J.; Van Oost, K. The interdisciplinary nature of SOIL. *Soil* **2015**, *1*, 117–129. [CrossRef]

25.	DeBano, L.F. The effect of hydrophobic substances on water movement in soil during infiltration. *Proc. Soil Sci. Soc. Am.* **1971**, *35*, 340–343. [CrossRef]

26.	Doerr, S.H.; Ferreira, A.J.D.; Walsh, R.P.D.; Shakesby, R.A.; Leighton-Boyce, G.; Coelho, C.O.A. Soil water repellency as a potential parameter in rainfall-runoff modelling: Experimental evidence at point to catchment scales from Portugal. *Hydrol. Process.* **2003**, *17*, 363–377. [CrossRef]

27.	Cerdà, A.; Doerr, S.H. Influence of vegetation recovery on soil hydrology and erodibility following fire: An 11-year investigation. *Int. J. Wildland Fire* **2005**, *14*, 423–437. [CrossRef]

28.	Eynard, A.; Schumacher, T.E.; Lindstrom, M.J.; Malo, D.D. Effects of agricultural management systems on soil organic carbon in aggregates of Ustolls and Usterts. *Soil Tillage Res.* **2005**, *81*, 253–263. [CrossRef]

29.	Wallis, M.G.; Horne, D.J. Soil water repellency. In *Advances in Soil Science*; Stewart, B.A., Ed.; Springer: New York, NY, USA, 1992; Volume 20, pp. 91–146.

30. Blanco-Canqui, H. Does no-till farming induce water repellency to soils? *Soil Use Manag.* **2011**, *27*, 2–9. [CrossRef]

31. Blanco-Canqui, H.; Lal, R. Extent of soil water repellency under long-term notill soils. *Geoderma* **2009**, *149*, 171–180. [CrossRef]

32. García-Moreno, J.; Gordillo-Rivero, Á.J.; Zavala, L.M.; Jordán, A.; Pereira, P. Mulch application in fruit orchards increases the persistence of soil water repellency during a 15-years period. *Soil Tillage Res.* **2013**, *130*, 62–68. [CrossRef]

33. Romero Diaz, A.; Barbera, G.G.; Lopez Bermudez, F. Relationship between soil erosion, rainfall and vegetation cover in the semiarid environment of south east of Iberian Peninsula. In Proceedings of the Conference on Erosion and Land Degradation in the Mediterranean, Aveiro, Portugal, 14–18 June 1995; pp. 59–73.

34. Vacca, A.; Loddo, S.; Ollesch, G.; Puddu, R.; Serra, G.; Tomasi, D.; Aru, A. Measurement of runoff and soil erosion in three areas under different land use in Sardinia (Italy). *CATENA* **2000**, *40*, 69–92. [CrossRef]

35. Kottek, M.; Grieser, J.; Beck, C.; Rudolf, B.; Rubel, F. World Map of the Köppen-Geiger climate classification updated. *Meteorologische Zeitschrift* **2006**, *15*, 259–263. [CrossRef]

36. FAO-UNESCO. *Soil Map of the World, 1:5,000,000*; UNESCO: Paris, France, 1981.

37. IUSS Working Group WRB. *World Reference Base for Soil Resources 2014. International Soil Classification System for Naming Soils and Creating Legends for Soil Maps. Update 2015*; World Soil Resources Report 106; FAO: Rome, Italy, 2015; p. 188.

38. Guitián, F.; Carballas, T. *Técnicas de análisis de suelos*; Ed. Pico Sacro: Santiago de Compostela, Spain, 1976.

39. Nelson, D.W.; Sommers, L.E. Total carbon, organic carbon, and organic matter. In *Methods of Soil Analysis, Part 3*; Sparks, D.L., Page, A.L., Helmke, P.A., Loeppert, R.H., Eds.; SSSA Book Series; American Society of Agronomy: Madison, WI, USA, 1996; pp. 961–1010.

40. Hedo, J.; Lucas-Borja, M.E.; Wic, B.; Andrés Abellán, M.; De las Heras, J. Experimental site and season over-control the effect of Pinus halepensis in microbial properties of soil under semiarid and dry conditions. *J. Arid Environ.* **2015**, *116*, 44–52. [CrossRef]

41. Bremmer, J.M.; Mulvaney, C.S. Nitrogen—Total. In *Methods of Soil Analysis*; Page, A.L., Miller, R.H., Keeney, D.R., Eds.; American Society of Agronomy: Madison, WI, USA, 1982; pp. 595–624.

42. Olsen, S.R.; Cole, C.V.; Watanabe, F.S.; Dean, L.A. *Estimation of Available Phosphorus in Soils by Extraction with Sodium Bicarbonate*; Circular 939; U.S. Department of Agriculture: Washington, DC, USA, 1954; pp. 1–19.

43. Thomas, G.W. Exchangeable Cations. In *Methods of Soil Analysis*; Page, A.L., Miller, R.H., Keeney, D.R., Eds.; American Society of Agronomy: Madison, WI, USA, 1982; pp. 159–165.

44. Decagon Devices. *Minidisk Infiltrometer User's Manual Version 10*; Decagon Devices: Pullman, WA, USA, 2012.

45. Zhang, R. Determination of soil sorptivity and hydraulic conductivity from the disk infiltrometer. *Soil Sci. Soc. Am. J.* **1997**, *61*, 1024–1030. [CrossRef]

46. Doerr, S.H. On standardizing the 'water drop penetration time' andthe 'molarity of an ethanol droplet' techniques to classify soil hydrophobicity: A case study using medium textured soils. *Earth Surf. Process. Landf.* **1998**, *23*, 663–668. [CrossRef]

47. Plaza-Álvarez, P.A.; Lucas-Borja, M.E.; Sagra, J.; Zema, D.A.; González-Romero, J.; Moya, D.; De las Heras, J. Changes in soil hydraulic conductivity after prescribed fires in Mediterranean pine forests. *J. Environ. Manag.* **2019**, *232*, 1021–1027. [CrossRef]

48. Zema, D.A.; Bombino, G.; Andiloro, S.; Zimbone, S.M. Irrigation of energy crops with urban wastewater: Effects on biomass yields, soils and heating values. *Agric. Water Manag.* **2012**, *115*, 55–65. [CrossRef]

49. Brady, N.C.; Weil, R.R. *The Nature and Properties of Soils*, 13th ed.; Prentice Hall: Upper Saddle River, NJ, USA, 2002.

50. Cerdà, A. Changes in overland flow and infiltration after a rangeland fire in a Mediterranean scrubland. *Hydrol. Process.* **1998**, *12*, 1031–1042. [CrossRef]

51. Cerdà, A. Postfire dynamics of erosional processes under Mediterranean climatic conditions. *Zeitschrift fur Geomorphologie* **1998**, *42*, 373–398.

52. Doerr, S.H.; Shakesby, R.A.; Dekker, L.W.; Ritsema, C.J. Occurrence, prediction and hydrological effects of water repellency amongst major soil and land-use types in a humid temperate climate. *Eur. J. Soil Sci.* **2006**, *57*, 741–754. [CrossRef]

53. DeBano, L.F. Water repellency in soils: A historical overview. *J. Hydrol.* **2000**, *231–232*, 4–32. [CrossRef]

54. Dekker, L.W.; Ritsema, C.J. Variation in water content and wetting patterns in Dutch water repellent peaty clay and clayey peat soils. *CATENA* **1996**, *28*, 89–105. [CrossRef]

55. Doerr, S.H.; Ritsema, C.J.; Dekker, L.W.; Scott, D.F.; Carter, D. Water repellence of soils: New insights and emerging research needs. *Hydrol. Process. Int. J.* **2007**, *21*, 2223–2228. [CrossRef]

56. Hallett, P.D.; Gaskin, R.E. An introduction to soil water repellency. In Proceedings of the 8th International Symposium on Adjuvants for Agrochemicals (ISAA2007), Columbus, OH, USA, 6–9 August 2007; International Society for Agrochemical Adjuvants: Wageningen, The Netherlands, 2007.

57. Woche, S.K.; Goebel, M.O.; Kirkham, M.B.; Horton, R.; Van der Ploeg, R.R.; Bachmann, J. Contact angle of soils as affected by depth texture and land management. *Eur. J. Soil Sci.* **2005**, *56*, 239–251. [CrossRef]

58. Mataix-Solera, J.; Doerr, S.H. Hydrophobicity and aggregate stability in calcareous topsoils from fire-affected pine forests in southeastern Spain. *Geoderma* **2004**, *118*, 77–88. [CrossRef]

59. Granged, A.J.P.; Zavala, L.M.; Jordàn, A.; Munoz-Rojas, M.; Mataix-Solera, J. Short-term effects of experimental fire for a soil under eucalyptus forest (SE Australia). *Geoderma* **2011**, *167–168*, 125–134. [CrossRef]

60. González-Peñaloza, F.A.; Cerdà, A.; Zavala, L.M.; Jordán, A.; Giménez-Morera, A.; Arcenegui, V. Do conservative agriculture practices increase soil water repellency? A case study in citrus-cropped soils. *Soil Tillage Res.* **2012**, *124*, 233–239. [CrossRef]

61. Hallett, P.D. A brief overview of the causes, impacts and amelioration of soil water repellency—A review. *Soil Water Res.* **2008**, *3*, 521–528. [CrossRef]

62. Dekker, L.W.; Doerr, S.H.; Oostindie, K.; Ziogas, A.K.; Ritsema, C.J. Water repellency and critical soil water conent in a dune sand. *Soil Sci. Soc. Am. J.* **2001**, *65*, 1667–1674. [CrossRef]

63. Verheijen, F.G.A.; Cammeraat, L.H. The association between three dominant shrub species and water repellent soils along a range of soil moisture contents in semi-arid Spain. *Hydrol. Process.* **2007**, *21*, 2310–2316. [CrossRef]

64. Cammeraat, E.L.; Cerdà, A.; Imeson, A.C. Ecohydrological adaptation of soils following land abandonment in a semi-arid environment. *Ecohydrology* **2010**, *3*, 421–430. [CrossRef]

65. Cerdà, A. The effect of patchy distribution of *Stipa tenacissima* L. on runoff and erosion. *J. Arid Environ.* **1997**, *36*, 37–51. [CrossRef]

66. Keesstra, S.; Pereira, P.; Novara, A.; Brevik, E.C.; Azorin-Molina, C.; Parras-Alcántara, L.; Jordán, A.; Cerdà, A. Effects of soil management techniques on soil water erosion in apricot orchards. *Sci. Total Environ.* **2016**, *551*, 357–366. [CrossRef] [PubMed]

67. Megahan, W.F.; Molitor, D.C. Erosional effects of wildfire and logging in Idaho. In Proceedings of the Symposium on Watershed Management, American Society of Civil Engineers, Cogan, UT, USA, 14–16 August 1975; pp. 423–444.

68. Prosser, I.P.; Williams, L. The effect of wildfire on runoff and erosion in native Eucalyptus forest. *Hydrol. Process.* **1998**, *12*, 251–265. [CrossRef]

69. Lesschen, J.P.; Cammeraat, L.H.; Kooijman, A.M.; van Wesemael, B. Development of spatial heterogeneity in vegetation and soil properties after land abandonment in a semi-arid ecosystem. *J. Arid Environ.* **2008**, *72*, 2082–2092. [CrossRef]

70. Moody, J.A.; Shakesby, R.A.; Robichaud, P.R.; Cannon, S.H.; Martin, D.A. Current research issues related to post-wildfire runoff and erosion processes. *Earth Sci. Rev.* **2013**, *122*, 10–37. [CrossRef]

71. Shakesby, R.A. Post-wildfire soil erosion in the Mediterranean: Review and future research directions. *Earth Sci. Rev.* **2011**, *105*, 71–100. [CrossRef]

 *water*

Article

Impact of Farmland Abandonment on Water Resources and Soil Conservation in Citrus Plantations in Eastern Spain

Artemi Cerdà [1], Oren Ackermann [2], Enric Terol [3] and Jesús Rodrigo-Comino [4,*]

[1] Soil Erosion and Degradation Research Group, Department of Geography, Valencia University, Blasco Ibàñez, 28, 46010 Valencia, Spain; artemio.cerda@uv.es

[2] Israel Heritage Department and the Department of Land of Israel Studies and Archaeology, Ariel University, P.O.B. 3, Ariel 4070000, Israel; orenac@ariel.ac.il

[3] Department of Cartographic Engineering, Geodesy and Photogrammetry, Universitat Politècnica de València, Camino de Vera, s/n, 46022 Valencia, Spain; eterol@cgf.upv.es

[4] Instituto de Geomorfología y Suelos, Department of Geography, University of Málaga, Edificio Ada Byron, Ampliación del Campus de Teatinos, 29071 Málaga, Spain

* Correspondence: rodrigo-comino@uma.es or geo.jrc@gmail.com

Received: 28 March 2019; Accepted: 17 April 2019; Published: 19 April 2019

Abstract: Due to the reduction in the prices of oranges on the market and social changes such as the ageing of the population, traditional orange plantation abandonment in the Mediterranean is taking place. Previous research on land abandonment impact on soil and water resources has focused on rainfed agriculture abandonment, but there is no research on irrigated land abandonment. In the Valencia Region—the largest producer of oranges in Europe—abandonment is resulting in a quick vegetation recovery and changes in soil properties, and then in water erosion. Therefore, we performed rainfall simulation experiments (0.28 m^2; 38.8 mm h^{-1}) to determine the soil losses in naveline orange plantations with different ages of abandonment (1, 2, 3, 5, 7 and 10 years of abandonment) which will allow for an understanding of the temporal changes in soil and water losses after abandonment. Moreover, these results were also compared with an active plantation (0). The results show that the soils of the active orange plantations have higher runoff discharges and higher erosion rates due to the use of herbicides than the plots after abandonment. Once the soil is abandoned for one year, the plant recovery reaches 33% of the cover and the erosion rate drops one order of magnitude. This is related to the delay in the runoff generation and the increase in infiltration rates. After 2, 3, 5, 7 and 10 years, the soil reduced bulk density, increase in organic matter, plant cover, and soil erosion rates were found negligible. We conclude that the abandonment of orange plantations reduces soil and water losses and can serve as a nature-based solution to restore the soil services, goods, and resources. The reduction in the soil losses was exponential (from 607.4 g m^{-2} in the active plot to 7.1 g m^{-2} in the 10-year abandoned one) but the water losses were linear (from 77.2 in active plantations till 12.8% in the 10-year abandoned ones).

Keywords: agricultural land management; irrigated fields; erosion; abandonment; soil properties

---

1. Introduction

Citrus plantations are widespread due to the increase in the fruit demand around the world [1,2]. The traditional citrus production regions such as Valencia face a crisis due to production in other regions [3]. A social and economic change is taking place in traditional citrus production areas due to the ageing of the population, the competition of the large farms managed by companies, and social and economic changes [4]. The small size of traditional farms is also a key concern for the viability of production [5]. Citrus production in Valencia is facing technical modernization in large farms, with drip irrigation and highly mechanized systems where labor is reduced [6] but the soil and water losses are enhanced (Keesstra et al., 2019). In front, we have traditional small size farms, with flood irrigation and high labor costs that result in low income for the farmers. Moreover, in the Valencia region, the ageing of the farmers and the social changes (to be a farmer is not attractive for the new generation) and low income associated with farming results in farmland abandonment [7].

Land abandonment has been widely researched in the last 50 years [8,9]. This is a consequence of the changes in the economy of developed countries rather than triggered different drivers [10,11]. From an environmental point of view, abandonment results in an increase in the infiltration rates [12] due to vegetation recovery [13]. Soil erosion is also reduced by the effect of the vegetation cover and the recovery of organic matter [14].

The research carried out on recent land abandonment was done in rainfed land. This is due to the fact that the abandonment took place in rainfed land that did not evolve to irrigation [15]. The intensification of the agriculture contributed to increasing the land irrigated; meanwhile, the one that was less productive, usually the rainfed one, was abandoned. The research carried out on irrigated land is recent and focuses mainly on socioeconomic issues [16]. The abandonment provides ecosystems services such as carbon sequestration, water storage, soil properties, and seed bank fate improvements [17]. This response to land abandonment is widespread around the world [18–20].

Research on the abandonment of traditional irrigation farms was restricted until now to the historical, archaeological, and ecological approach [21–23]. There is no information about the changes in soil hydrology and soil erosion once the abandonment takes place in traditional flood irrigated land. This paper is the first approach to understand the effect of irrigated farmland abandonment on soil erosion and runoff discharge. Our main goal is to determine how the soil and water resources change once abandonment takes place and how to evolve along a decade with the use of plots in use or abandoned for 1, 2, 3, 5, 7 and 10 years.

2. Materials and Methods

2.1. Study Area

The research presented in this paper was carried out in the Canyoles River watershed, in the fluvial terraces (145 m a.s.l.) in the municipality of Canals (Valencia, Spain), a representative zone of the Mediterranean citrus plantation that was established along the 1960s. The farm properties are small in size in this region, and usually, all the farms are smaller than 1 ha and the average at the study site is 0.45 ha. The active field is flood irrigated and herbicide (Glyphosate) is applied any time the weeds germinate, which results in a bare soil surface. Once abandonment takes place there is a quick recovery of the vegetation. The climate is typically Mediterranean with a mean annual precipitation of 532 mm and a mean annual temperature of 16.2 °C. The soil is an Anthrosol [24] and the texture of the soil is clay-loam: 21% clay, 39% silt, and 40% sand. The planting pattern is 5 m × 4 m, the usual pattern for citrus in this agricultural area. The farm was flood-irrigated with freshwater from the Riu de Sants, which flows from the Massís del Caroig aquifer. The slope is negligible due to the land flattering and the effects of the floods on the fields. The soils are basic in pH (7.9). The observed soil profiles were very homogeneous as a consequence of tillage practices for centuries and land levelling. The study area was selected in the Pleistocene fluvial terrace of the Canyoles River (near the hamlet of Aiacor) and show 22% gravel content. The irrigation system (Sèquia de Ranes) flows from the nearby Riu de Sants spring. Irrigation takes place every 2 weeks in the summer and does not take place once the fields are abandoned. The management up until the time of abandonment was similar in all plots: herbicide (Glyphosate (N-(phosphonomethyl)glycine) and inorganic fertilizers (NPK—nitrogen, phosphorus, and potassium—0.8 Mg ha−1 per year). Once the plots are abandoned, neither irrigation nor fertilization takes place.

2.2. Experimental Design

The experimental design was based on the fact that the study area has suffered since 2000 a progressive abandonment due to the low income in such small parcels, the dropping of the prices of the oranges, and the ageing of the landowners. Then we selected a farm that was active and the neighbouring farms that had been abandoned for 1, 2, 3, 5, 7 and 10 years (Figure 1A). At each farm, we selected 10 representative 1-m^2 plot where sampling and simulated rainfall experiments on 0.28-m^2 plots were carried out (Figure 1B). All the plots have *Citrus sinensis* plantations with Naveline variety and "Carrizo" rootstock.

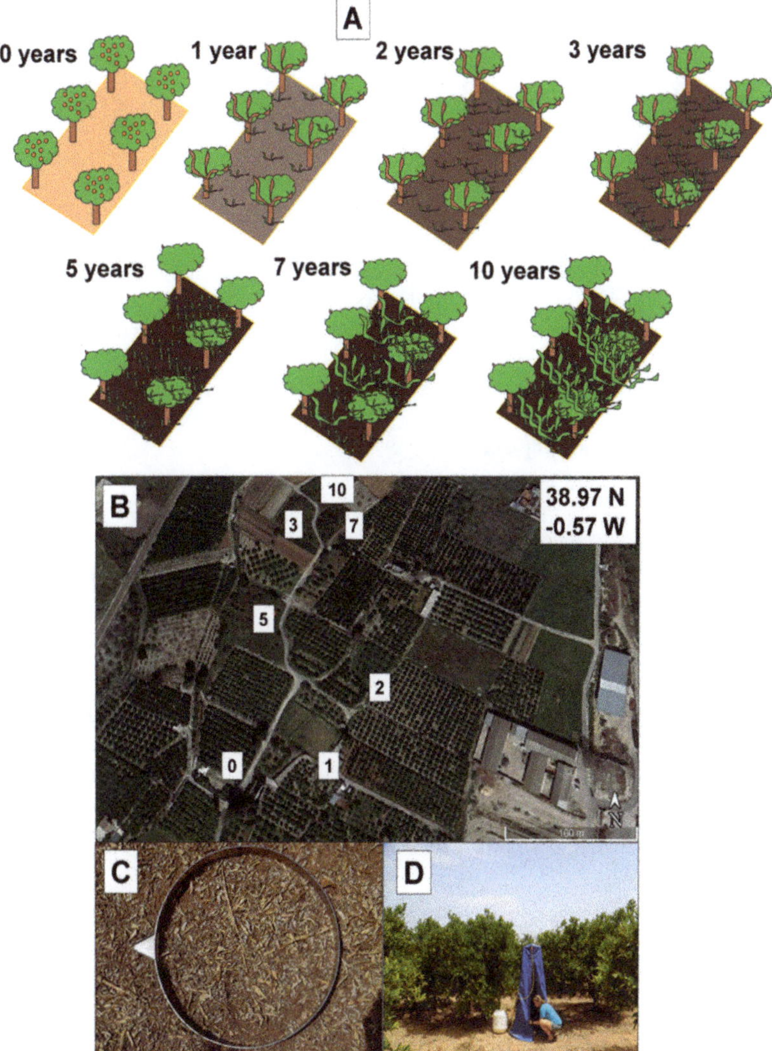

Figure 1. Scheme of the study plots (**A**), localization of the experimental fields (**B**), ring plot in the active farm (**C**), and field campaign (**D**).

2.3. Soil Sampling

The seven experimental fields show different ages of abandonment. Age 0 is the active field with no weed cover due to the use of herbicides, and 1, 2, 3, 5, 7, and 10 years are the fields that were abandoned along the last decade. The soil sampling took place in August 2014 during the Mediterranean summer drought. Ten soil samples were taken from the top 6 cm of the soil using a 100 cm^3 steel cylinder. We measured gravimetric soil water content (SWC) and bulk density (BD). Moreover, soil samples were weighed, oven dried at 105 °C during 24 h, and weighed at room temperature to estimate organic matter by the dichromate method [25]. Also, grain size distribution was calculated by the pipette method. Vegetation cover, stones, and soil crusts were estimated using 100 pins evenly distributed in the rainfall simulation plot.

2.4. Rainfall Simulation Experiments

Seventy experiments (10 repetitions × 7 plots) during 1 h for the active and 1, 2, 3, 5, 7, and 10 years of abandonment were carried out at 38.8 mm h^{-1} rainfall intensity on the circular paired ring plot (0.28 m^2; Figure 1C,D). These rainfall events are characterised by a return period around 2 years in the eastern Iberian Peninsula and have been used widely in rainfall simulation experiments [26]. All experiments were carried out when the soil moisture was low during the last week of July 2014 with no rainfall events to avoid variability among plots. Runoff was collected each 1-min interval and total water loss was calculated. Then, the runoff coefficient was obtained by means of the percentage of rainfall water flowing through the ring plot. In the laboratory, runoff samples were desiccated (105 °C, 24 h) and soil loss was obtained based on the weight basis per area and time (g m^2 h^{-1}). Furthermore, it was also possible to quantify time to ponding (time required for 40% of the surface to be ponded; Tp, s), time to runoff initiation (Tr, s), and time required by runoff to reach the outlet (Tro, s). Those parameters show how the runoff initiation takes place and how ponding is transformed into runoff and how the runoff reaches the runoff outlet.

2.5. Statistical Analysis

Descriptive statistics were calculated (mean, median, maximum, and minimum) and used for further statistical analysis. Soil properties were represented in linear graphs in order to show their evolutions during the studied period using Excel software (Windows, Redmond, Washington, DC, USA). The number of points used for each analysis corresponded to all measurements done at each location. Then, the hydrological response was represented as polar graphs dividing by intervals (natural breaks) the obtained results. Finally, soil erosion results were depicted in scatterplots using SigmaPlot 12.0 (Systat, Chicago, IL, USA). Statistical differences were evaluated performing an ANOVA one-way test to check the statistically significant differences among years of abandonment. If the normality test failed (Shapiro–Wilk), a Tukey test was conducted when the homogeneity variance fails (Levene´s test). Finally, a Spearman rank correlation coefficient was carried out in order to assess the possible connection between environmental plot characteristics and soil erosion results.

3. Results

3.1. Soil Properties

The soils of the seven study sites affected by different times since abandonment show changes that can shed light into the evolution of the soil properties upon abandonment (Figure 2). Antecedent soil moisture is reduced from 14% to 5.1% in ten years of abandonment. Bulk density does not show significative changes after 1 year of abandonment (from 1.38 to 1.37 g cm^{-3}), but after that, the values decrease to 1.15 g cm^{-3} in the tenth year of abandonment. On the contrary, organic matter and plant cover quickly increased after the abandonment, showing changes from 0.93 to 1.79% and from 1.1 to 90.3%, respectively. The cover of rock fragments varies among plots, reaching the maximum values in the plots after 1, 2, and 3 years of abandonment. Finally, the development of soil crusts also shows a drastic decrease from the active plot (83.4%) to the 10 years abandoned plot (3.4%).

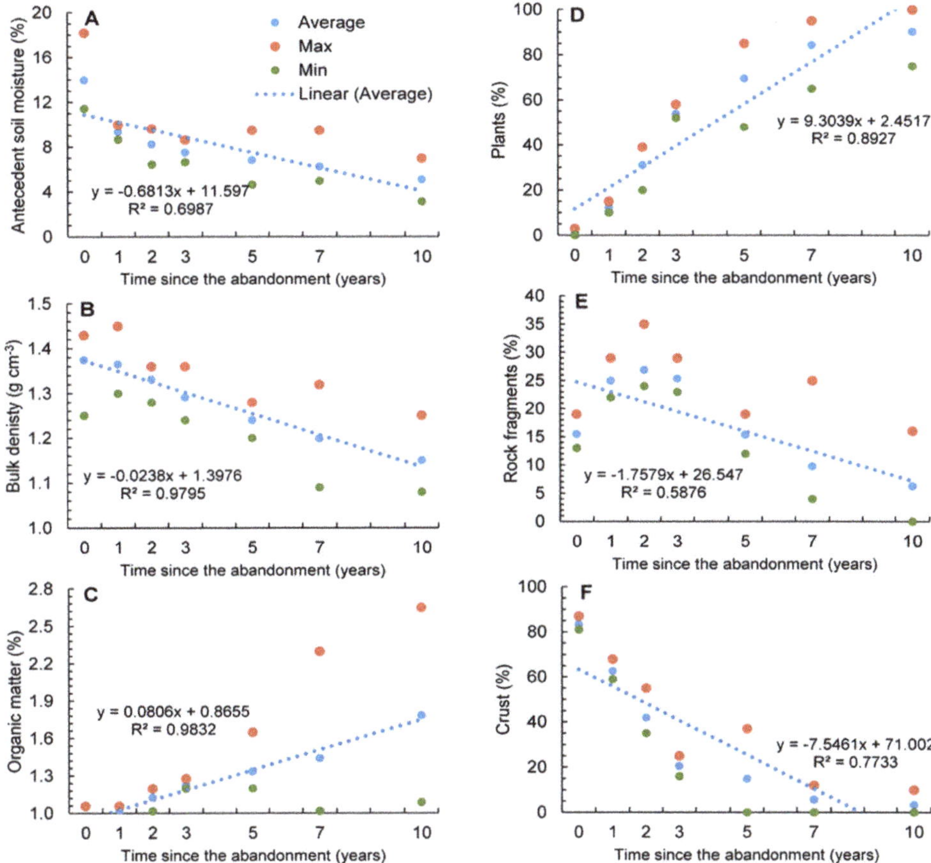

Figure 2. Variation of soil properties over different time periods since abandonment. (**A**) Antecentt soil moisture; (**B**); Bulk density; (**C**); Organic matter; (**D**); Plants; (**E**) Rock fragments; (**F**) Crust.

3.2. Hydrological Response

Figure 3 shows the time to ponding, time to runoff generation, and time to runoff in the outlet (seconds) represented in polar graphs. Table 1 shows the results of the Tukey test applied to all the seven study sites. The results show that among plots, there exists a clear statistically significant difference after the abandonment ($p < 0.001$). Time to ponding moves from 25.9 to 238.1 seconds, respectively, from the active plot to the 10 years abandoned one. Time to ponding was much delayed after seven years of abandonment. Runoff was generated after 47.2 seconds in the cultivated plot and after the runoff was delayed by 477.8 seconds. There was a clear trend that showed a progressive delay in the ponding and runoff initiation once the abandonment took place. Moreover, a clear variance among the 10 repetitions in each plot was also relevant, which could be affected by the uneven distribution of the plant recovery. The time to a runoff in the outlet was faster (88.3 s) in the active plot and delayed after the abandoned. The 10-year abandoned plot needed 917 s to achieve runoff at the outlet.

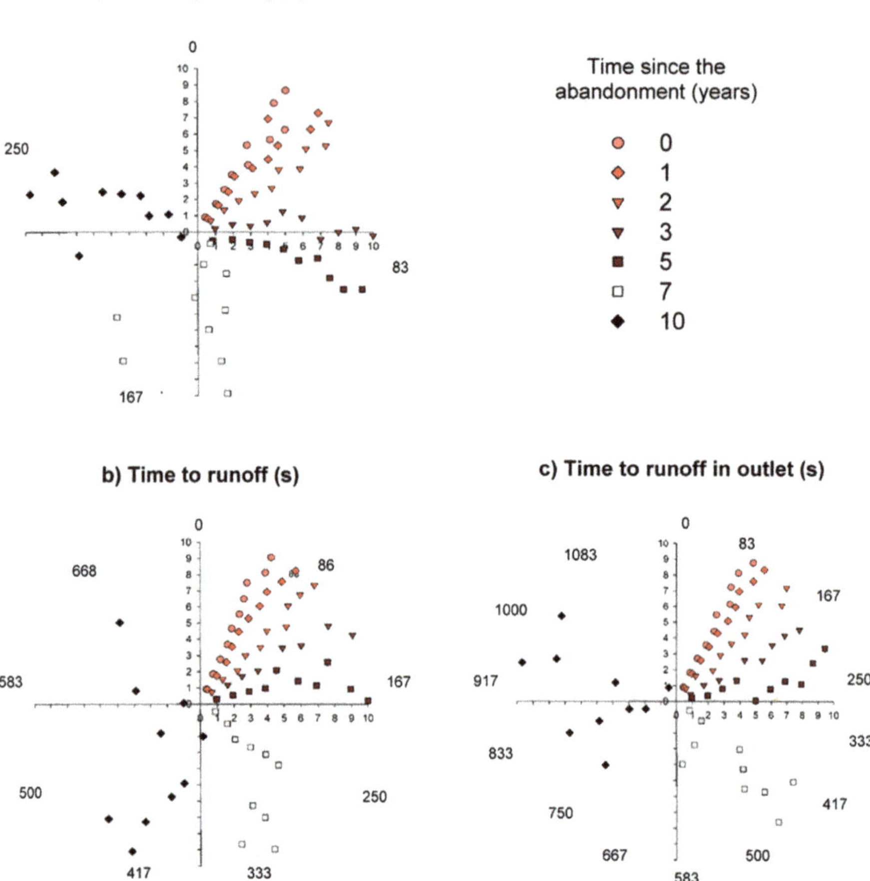

Figure 3. Hydrological response depending on the period of abandonment (360° = range of values).

Table 1. Statistical differences of hydrological responses and soil erosion results among periods of abandonment. Tp: time to ponding; Tr: time to runoff generation; Tro: time to runoff in outlet; Rc: runoff coefficient; SC: sediment concentration; SL: soil loss.

Years	Tp	Tr	Tro	Rc	SL	SC
0 vs. 1	<0.001	<0.001	<0.001	**0.536**	0.005	<0.001
0 vs. 2	<0.001	<0.001	<0.001	0.007	<0.001	<0.001
0 vs. 3	<0.001	<0.001	<0.001	<0.001	<0.001	<0.001
0 vs. 5	<0.001	<0.001	<0.001	<0.001	<0.001	<0.001
0 vs. 7	<0.001	<0.001	<0.001	<0.001	<0.001	<0.001
0 vs. 10	<0.001	<0.001	<0.001	<0.001	<0.001	<0.001
1 vs. 2	<0.001	<0.001	<0.001	0.036	<0.001	0.338
1 vs. 3	<0.001	<0.001	<0.001	<0.001	<0.001	<0.001
1 vs. 5	<0.001	<0.001	<0.001	<0.001	<0.001	<0.001
1 vs. 7	<0.001	<0.001	<0.001	<0.001	<0.001	<0.001
1 vs. 10	<0.001	<0.001	<0.001	<0.001	<0.001	<0.001
2 vs. 3	<0.001	<0.001	<0.001	0.205	<0.001	0.002
2 vs. 5	<0.001	<0.001	<0.001	0.014	<0.001	<0.001
2 vs. 7	<0.001	<0.001	<0.001	<0.001	<0.001	<0.001
2 vs. 10	<0.001	<0.001	<0.001	<0.001	<0.001	<0.001
3 vs. 5	0.002	<0.001	<0.001	<0.001	0.002	0.048
3 vs. 7	<0.001	<0.001	<0.001	<0.001	<0.001	0.014
3 vs. 10	<0.001	<0.001	<0.001	<0.001	<0.001	<0.001
5 vs. 7	<0.001	<0.001	<0.001	<0.001	<0.001	0.400
5 vs. 10	<0.001	<0.001	<0.001	<0.001	<0.001	<0.001
7 vs. 10	<0.001	<0.001	<0.001	0.005	<0.001	0.003

3.3. Runoff, Soil Loss, and Sediment Concentration

In Figure 4, runoff coefficient, sediment concentration, and soil loss are depicted in box plots. The runoff coefficient shows the highest rates in the active plot, reaching average values of 77.2% of the rainfall, with maximum ones of 96.3%. On the contrary, in the other plots, the average runoff coefficient descends to 71.8, 51, 42.3, 32.3, 22.3, and 12.8% for 1, 2, 3, 5, 7- and 10-years of abandonment, respectively. The highest variability can be observed after the second year of abandonment. In Table 1, it is possible to observe values with non-statistically significant differences in runoff coefficients between 0–1 year and 2–3-years after abandonment. Sediment concentration also shows a drastic decrease from the active plot to the 10-year abandoned field, obtaining values from 14.7 (maximum values of 20.4 g L^{-1}) to 1 g L^{-1} (maximum values of 1.02 g L^{-1}). Table 2 shows that all the plots have significant differences in this parameter. The soil loss parameter also shows a decrease: from 607.4 (active plot), 271.4 (1 year), 150.3 (2 years), 62.8 (3 years), 33.46 (5 years), 19.7 (7 years) till 7.1 g m^{-2} (10 years). Table 1 shows that between 1–2 years and 5–7 years of abandonment, no differences can be observed, but there are differences in other years.

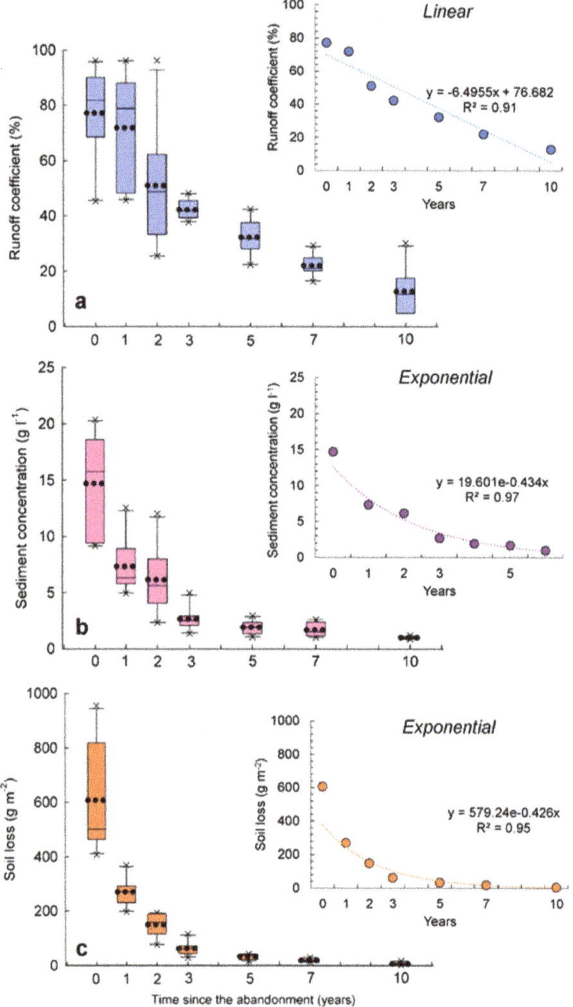

Figure 4. Evolution of the runoff coefficient, sediment concentration, and soil loss along the abandonment time from active plantations to 10-years old abandonment. Dotted line: average (n = 10). (**a**) Runoff coefficient; (**b**) Sediment concentration; (**c**) Soil loss.

Table 2. Spearman rank coefficient between environmental factors, hydrological responses, and soil erosion. SM: antecedent soil moisture; BD: bulk density; OM: organic matter; Plants: plant cover; Rock: rock fragment cover; Crust: soil crust; Tp: time to ponding; Tr: time to runoff generation; Tro: time to runoff in outlet; Rc: runoff coefficient; SC: sediment concentration; SL: soil loss.

Variables	Tp	Tr	Tro	Rc	SC	SL
SM	−0.84	−0.84	−0.85	0.77	0.80	0.77
BD	−0.85	−0.86	−0.85	0.80	0.81	0.75
OM	0.82	0.83	0.83	−0.77	−0.76	−0.70
Plants	0.95	0.95	0.95	−0.87	−0.88	−0.84
Rock	−0.56	−0.54	−0.54	0.59	0.49	0.49
Crust	−0.93	−0.93	−0.93	0.85	0.85	0.81

In Table 2, it can be observed that only the rock fragment cover does not show a high correlation with the hydrological responses and soil erosion results. On the other hand, the highest Spearman rank coefficient values were obtained between the development of a high plant cover and the positive relationship with Tp (0.95), Tr (0.95), and Tro (0.95), and the negative one with Rc (−0.87), SSC (−0.88), and SL (−0.84). Moreover, high positive correlations were observed with the OM contents. On the contrary, higher values of antecedent soil moisture, bulk density, and soil crust were highly correlated with Tp, Tro, and Tro, and Rc, Sc, and SL.

4. Discussion

The use of developing countries as food producers, an increase in agricultural productivity, and the use of fossil fuels that reduce the need for animal traction and then for food production for them has reduced the need to use part of the agriculture land [27,28]. This abandonment has environmental consequences such as changes in water resources, vegetation cover and floristic composition, fauna, and soil properties, which will affect also earth system functioning as they will control the biogeochemical cycles [29–33]. Agriculture land abandonment is not a new issue from a social, economic, and environmental point of view [34], but what is new is the agriculture technology that was developed after War World II which has led to the widespread abandonment and contributed to the restoration of the vegetation and fauna and the reduction in the soil losses in many parts of the world [35,36].

The research on the impact of abandonment of agriculture on environmental issues focused on rainfed land [37], as the irrigated land used to be intensified [38,39]. Our research contributes to new data generated in flood irrigated land that is being abandoned due to socioeconomic changes. The findings demonstrated that the soils are able to recover plant cover and soil organic matter once abandoned as other authors demonstrated in other types of agricultural land [17,40]. The changes in plant cover will affect fauna such as has been demonstrated by other colleagues [41–43] and soil properties. The non-utilisation of irrigation has generated soils which contain a lower amount of water (antecedent soil moisture), which could affect some other pedological processes such as organic matter development or aggregate stability [44,45]. Considering these changes, the diminution of soil bulk density is also a relevant factor that could affect soil erosion activation [46]. As Bienes et al. [47] observed in Central Spain, significant changes in bulk density used to appear after the abandonment took place due to plant recolonization because of the root development and organic matter increase. In our study area, weeds, grass, and small shrubs were distributed in little patches that directly affect these pedological changes.

The plant cover (weeds) was mainly *Paretaria Judaica*, *Conyza sumatrensis*, *Amarathus hibridus* L., and *Chenopodium album*. Once abandoned, the cover of *Paretaria Judaica* increased, but other plants appeared such as *Avena fatua*, *Asparragus* sp., and *Rubus ulmifolius* that finally became the dominant species.

Once the land was abandoned, we also observed a significant difference in the rock fragment cover. Possibly, the surface wash and tillage could remove the fine material and more rock fragments were visible on the surface [48,49]. However, after the third year of abandonment, the vegetation cover reduced neither the crust nor the rock fragments in the soil surface. This would also be an interesting topic for research in the future because of its direct impacts on soil erosion processes and biogeological systems.

Our research confirms that after the abandonment there was a sudden reduction in overland flows that is shown in delayed ponding and runoff generation. This means that more water infiltrates. However, our measurements also demonstrate that the amount of water flowing through and on the soil is also reduced after the abandonment takes place. This can be due to the reduction in the effective rainfall as the interception increases with the growth of the vegetation and the development of a litter layer [50,51]. The impact of the litter was researched by the pioneer study of Facelli and Pickett [52] and found that litter plays a key role in the water balance. Rainfall interception in abandoned fields is related to the vegetation recovery, and the higher the biomass means a lower the net precipitation on the soil [53].

The loss of water reaching the soil after the abandonment of land changes the hydrological cycle, such as Hou et al. [54] found in the Loess hillslopes in China, Šraj et al. [50] in Slovenia, and Otero et al. [55] in Catalonia (Spain), where a loss of stream biodiversity and water availability was found. Those findings are highlighted by García-Ruiz and Lana-Renault [13] along with their review on the hydrological and erosive consequences of farmland abandonment in Europe. The impact of abandonment shows less water availability and more water used by the vegetation. This results in a river discharge reduction such as was demonstrated in the Central Spanish Pyrenees by Beguería et al. [56] or in Slovenia where the Dragonja River reduced the sediment delivery due to the loss of the runoff discharge as a consequence of the natural afforestation [57]. The use of water by the vegetation (transpiration) also resulted in the loss of water available by plants such as Moreira et al. [58] found in the Amazon on abandoned pastures, Rambousková et al. [59] in the abandoned fields in the Bohemian Karst, and Farrick and Price [60] in the Sphagnum restoration in peatlands.

Our research at the Canals municipality demonstrated that the reduction in the water yield was efficient as a consequence of the abandonment (from 77.2% in the active orange plantation to 12.8% in the 10-years abandoned plots), but the reduction shows a linear correlation (0.91; Figure 4a). However, for the sediment concentration and soil loss, the reduction followed a negative exponential trend with an adjustment of 0.97 and 0.95, respectively (Figure 4b,c). This trend in the reduction of soil and water losses were also found in previous literature. Ruecker et al. [61]; Koulouri and Giourga [62], and Lesschen et al. [63] found this trend in abandoned Mediterranean terraces, Cerdà [12] in southeastern Spain in a semi-arid landscape, Liu et al. [64] in the Loess Plateau, and Arnáez et al. [65] in the La Rioja region, wherein the Camero Viejo district they found that abandonment controlled the soil erosion rates and landscape evolution. Land abandonment is a recurrent topic in the Mediterranean, and Portuguese examples confirm our findings here: control of soil losses by vegetation recovery [66].

Abandonment could be seen as a nature-based solution such as the use of straw in forest fire affected land [66,67] to the high erosion rates found in agricultural land [68], and could be used as a strategy to balance the carbon cycle [69] and rehabilitate the soils under the millennia-old use of agriculture [70]. The research carried out at the Canals Municipality shows that abandonment could be positive from an environmental point of view, but there is also the risk of a forest fire as the vegetation could be very flammable during the Mediterranean summer drought [71]. Thus, the next essential step of research is to find the optimal management of the abandoned orchards, and maybe it can be used as a recreation area for nearby urban citizens [72]. This could be a successful solution in the Mediterranean where agriculture land abandonment in the last decade also takes place in the surroundings of city areas due to the economic crisis [69] and for which similar responses have been shown in other developed countries [73]. How urban changes occur is related in one way to the environmental conditions [74] but also to the cultural impact of humans [75].

Once the fields are abandoned, the vegetation recovery takes places as a mosaic of plants and this response results in an increase in the spatial variability such as other authors found along climatological gradients [76]. We found this increase in the spatial distribution due to the formation of tussocks, or bare and plant covered patches, which is a clear factor in soil erosion, but with more intensity after the second year of abandonment [77]. Our findings are based on a local survey and should be used to supply basic information to develop proper models of water balance and soil erosion [78] that will shed light onto the effect of other management systems such as organic farming, land abandonment or cover crops and mulches [79,80].

5. Conclusions

Citrus plantation abandonment results in a recovery of the vegetation cover and soil organic matter, and a reduction in the soil bulk density drought and soil moisture. Plant recovery is the key factor that determined a linear reduction in the runoff discharge (from 72 till 13% of the rainfall) over ten years of abandonment. The soil losses dropped from 607.4 g m^{-2} in the active plot to 7.1 g m^{-2} in the 10 years after abandonment took place. We conclude that the abandonment of orange plantations could reduce soil and water losses if there is a proper plant recovery, which allows it to be considered as a potential nature-based solution to restore the soil services, goods, and resources.

Author Contributions: A.C.: investigation, methodology, data acquisition, data preparation, writing, supervising. O.A.: data analysis, data preparation, writing. E.T.: investigation, methodology, data acquisition, data preparation, writing. J.R.C.: data analysis, data preparation, methodology, writing.

Funding: This paper is part of the results of research projects GL2008-02879/BTE, LEDDRA 243857 and RECARE-FP7 (ENV.2013.6.2-4).

Acknowledgments: During the field work the music of Raimon "Diguem no" was our inspiration and during the data analysis and writing the histories of "El Comandante Lara". Moreover, we want to thank the guest editors and reviewers for their careful review process.

Conflicts of Interest: The authors declare no conflict of interest.

References

1. Regmi, A. *Changing Structure of Global Food Consumption and Trade*; Department of Agriculture, Agriculture and Trade Report, Market and Trade Economics Division, Economic Research Service: Washington, DC, USA, 2001; ISBN 978-1-4289-4047-5.

2. Stefler, D.; Pikhart, H.; Kubinova, R.; Pajak, A.; Stepaniak, U.; Malyutina, S.; Simonova, G.; Peasey, A.; Marmot, M.G.; Bobak, M. Fruit and vegetable consumption and mortality in Eastern Europe: Longitudinal results from the Health, Alcohol and Psychosocial Factors in Eastern Europe study. *Eur. J. Prev. Cardiol.* **2016**, *23*, 493–501. [CrossRef] [PubMed]

3. Alford, M.; Barrientos, S.; Visser, M. Multi-scalar Labour Agency in Global Production Networks: Contestation and Crisis in the South African Fruit Sector. *Dev. Chang.* **2017**, *48*, 721–745. [CrossRef]

4. Cerdà, A.; Rodrigo-Comino, J.; Giménez-Morera, A.; Novara, A.; Pulido, M.; Kapović-Solomun, M.; Keesstra, S.D. Policies can help to apply successful strategies to control soil and water losses. The case of chipped pruned branches (CPB) in Mediterranean citrus plantations. *Land Use Policy* **2018**, *75*, 734–745. [CrossRef]

5. Ortega-Reig, M.; Sanchis-Ibor, C.; Palau-Salvador, G.; García-Mollá, M.; Avellá-Reus, L. Institutional and management implications of drip irrigation introduction in collective irrigation systems in Spain. *Agric. Water Manag.* **2017**, *187*, 164–172. [CrossRef]

6. Cerdà, A.; Rodrigo-Comino, J.; Giménez-Morera, A.; Keesstra, S.D. An economic, perception and biophysical approach to the use of oat straw as mulch in Mediterranean rainfed agriculture land. *Ecol. Eng.* **2017**, *108*, 162–171. [CrossRef]

7. Keesstra, S.D.; Rodrigo-Comino, J.; Novara, A.; Giménez-Morera, A.; Pulido, M.; Di Prima, S.; Cerdà, A. Straw mulch as a sustainable solution to decrease runoff and erosion in glyphosate-treated clementine plantations in Eastern Spain. An assessment using rainfall simulation experiments. *CATENA* **2019**, *174*, 95–103. [CrossRef]

8. Levers, C.; Schneider, M.; Prishchepov, A.V.; Estel, S.; Kuemmerle, T. Spatial variation in determinants of agricultural land abandonment in Europe. *Sci. Total Environ.* **2018**, *644*, 95–111. [CrossRef]

9. Kou, M.; Jiao, J.; Yin, Q.; Wang, N.; Wang, Z.; Li, Y.; Yu, W.; Wei, Y.; Yan, F.; Cao, B. Successional trajectory over 10 years of vegetation restoration of abandoned slope croplands in the Hill-Gully region of the Loess Plateau. *Land Degrad. Develop.* **2016**, *27*, 919–932. [CrossRef]

10. Ito, J.; Nishikori, M.; Toyoshi, M.; Feuer, H.N. The contribution of land exchange institutions and markets in countering farmland abandonment in Japan. *Land Use Policy* **2016**, *57*, 582–593. [CrossRef]

11. Lasanta, T.; Arnáez, J.; Pascual, N.; Ruiz-Flaño, P.; Errea, M.P.; Lana-Renault, N. Space–time process and drivers of land abandonment in Europe. *CATENA* **2017**, *149*, 810–823. [CrossRef]

12. Cerdà, A. Soil erosion after land abandonment in a semiarid environment of southeastern Spain. *Arid Soil Res. Rehabil.* **1997**, *11*, 163–176. [CrossRef]

13. García-Ruiz, J.M.; Lana-Renault, N. Hydrological and erosive consequences of farmland abandonment in Europe, with special reference to the Mediterranean region—A review. *Agric. Ecosyst. Environ.* **2011**, *140*, 317–338. [CrossRef]

14. Cerdà, A.; Rodrigo-Comino, J.; Novara, A.; Brevik, E.C.; Vaezi, A.R.; Pulido, M.; Giménez-Morera, A.; Keesstra, S.D. Long-term impact of rainfed agricultural land abandonment on soil erosion in the Western Mediterranean basin. *Prog. Phys. Geogr. Earth Environ.* **2018**, *42*, 202–219. [CrossRef]

15. Rodrigo-Comino, J.; Martínez-Hernández, C.; Iserloh, T.; Cerdà, A. Contrasted Impact of Land Abandonment on Soil Erosion in Mediterranean Agriculture Fields. *Pedosphere* **2018**, *28*, 617–631. [CrossRef]

16. Vidal-Macua, J.J.; Ninyerola, M.; Zabala, A.; Domingo-Marimon, C.; Gonzalez-Guerrero, O.; Pons, X. Environmental and socioeconomic factors of abandonment of rainfed and irrigated crops in northeast Spain. *Appl. Geogr.* **2018**, *90*, 155–174. [CrossRef]

17. Alonso-Sarría, F.; Martínez-Hernández, C.; Romero-Díaz, A.; Cánovas-García, F.; Gomariz-Castillo, F. Main environmental features leading to recent land abandonment in Murcia Region (Southeast Spain). *Land Degrad. Dev.* **2016**, *27*, 654–670. [CrossRef]

18. Gispert, M.; Pardini, G.; Colldecarrera, M.; Emran, M.; Doni, S. Water erosion and soil properties patterns along selected rainfall events in cultivated and abandoned terraced fields under renaturalisation. *CATENA* **2017**, *155*, 114–126. [CrossRef]

19. Horel, Á.; Tóth, E.; Gelybó, G.; Kása, I.; Bakacsi, Z.; Farkas, C. Effects of Land Use and Management on Soil Hydraulic Properties. *Open Geosci.* **2015**, *7*, 742–754. [CrossRef]

20. Yu, W.J.; Jiao, J.Y.; Chen, Y.; Wang, D.L.; Wang, N.; Zhao, H.K. Seed Removal due to Overland Flow on Abandoned Slopes in the Chinese Hilly Gullied Loess Plateau Region. *Land Degrad. Dev.* **2017**, *28*, 274–282. [CrossRef]

21. Löw, F.; Fliemann, E.; Abdullaev, I.; Conrad, C.; Lamers, J.P.A. Mapping abandoned agricultural land in Kyzyl-Orda, Kazakhstan using satellite remote sensing. *Appl. Geogr.* **2015**, *62*, 377–390. [CrossRef]

22. Lightfoot, D.R. Syrian qanat Romani: History, ecology, abandonment. *J. Arid Environ.* **1996**, *33*, 321–336. [CrossRef]

23. Verheyen, K.; Bossuyt, B.; Hermy, M.; Tack, G. The land use history (1278–1990) of a mixed hardwood forest in western Belgium and its relationship with chemical soil characteristics. *J. Biogeogr.* **1999**, *26*, 1115–1128. [CrossRef]

24. IUSS Working Group WRB. *World Reference Base for Soil Resources 2014*; World Soil Resources Report; FAO: Rome, Italy, 2014.

25. Walkley, A.; Black, I.A. An examination of Degtjareff method for determining soil organic matter and a proposed modification of the chromic acid titration method. *Soil Sci.* **1934**, *37*, 29–37. [CrossRef]

26. Iserloh, T.; Fister, W.; Seeger, M.; Willger, H.; Ries, J.B. A small portable rainfall simulator for reproducible experiments on soil erosion. *Soil Tillage Res.* **2012**, *124*, 131–137. [CrossRef]

27. Grădinaru, S.R.; Kienast, F.; Psomas, A. Using multi-seasonal Landsat imagery for rapid identification of abandoned land in areas affected by urban sprawl. *Ecol. Indic.* **2019**, *96*, 79–86. [CrossRef]

28. Basualdo, M.; Huykman, N.; Volante, J.N.; Paruelo, J.M.; Piñeiro, G. Lost forever? Ecosystem functional changes occurring after agricultural abandonment and forest recovery in the semiarid Chaco forests. *Sci. Total Environ.* **2019**, *650*, 1537–1546. [CrossRef] [PubMed]

29. Yin, H.; Prishchepov, A.V.; Kuemmerle, T.; Bleyhl, B.; Buchner, J.; Radeloff, V.C. Mapping agricultural land abandonment from spatial and temporal segmentation of Landsat time series. *Remote Sens. Environ.* **2018**, *210*, 12–24. [CrossRef]

30. Yin, H.; Butsic, V.; Buchner, J.; Kuemmerle, T.; Prishchepov, A.V.; Baumann, M.; Bragina, E.V.; Sayadyan, H.; Radeloff, V.C. Agricultural abandonment and re-cultivation during and after the Chechen Wars in the northern Caucasus. *Glob. Environ. Chang.* **2019**, *55*, 149–159. [CrossRef]

31. Baba, Y.G.; Tanaka, K.; Kusumoto, Y. Changes in spider diversity and community structure along abandonment and vegetation succession in rice paddy ecosystems. *Ecol. Eng.* **2019**, *127*, 235–244. [CrossRef]

32. Klee, R.J.; Zimmerman, K.I.; Daneshgar, P.P. Community Succession after Cranberry Bog Abandonment in the New Jersey Pinelands. *Wetlands* **2019**, *39*. [CrossRef]

33. Agnoletti, M.; Errico, A.; Santoro, A.; Dani, A.; Preti, F. Terraced Landscapes and Hydrogeological Risk. Effects of Land Abandonment in Cinque Terre (Italy) during Severe Rainfall Events. *Sustainability* **2019**, *11*, 235. [CrossRef]

34. Fredh, E.D.; Lagerås, P.; Mazier, F.; Björkman, L.; Lindbladh, M.; Broström, A. Farm establishment, abandonment and agricultural practices during the last 1300 years: A case study from southern Sweden based on pollen records and the LOVE model. *Veget. Hist. Archaeobot.* **2019**, *39*. [CrossRef]

35. Lasanta, T.; Nadal-Romero, E.; Arnáez, J. Managing abandoned farmland to control the impact of re-vegetation on the environment. The state of the art in Europe. *Environ. Sci. Policy* **2015**, *52*, 99–109. [CrossRef]

36. Bell, S.; Alves, S.; Silveirinha de Oliveira, E.; Zuin, A. Migration and land use change in Europe: A review. *Living Rev. Landsc. Res.* **2010**, *4*. [CrossRef]

37. Rodrigo-Comino, J.; Senciales, J.M.; Sillero-Medina, J.A.; Gyasi-Agyei, Y.; Ruiz-Sinoga, J.D.; Ries, J.B. Analysis of Weather-Type-Induced Soil Erosion in Cultivated and Poorly Managed Abandoned Sloping Vineyards in the Axarquía Region (Málaga, Spain). *Air Soil Water Res.* **2019**, *12*, 1178622119839403. [CrossRef]

38. Carter, D.L. Furrow Irrigation Erosion Lowers Soil Productivity. *J. Irrig. Drain. Eng.* **1993**, *119*, 964–974. [CrossRef]

39. Al-Ghobari, H.M.; Dewidar, A.Z. Integrating deficit irrigation into surface and subsurface drip irrigation as a strategy to save water in arid regions. *Agric. Water Manag.* **2018**, *209*, 55–61. [CrossRef]

40. Cammeraat, E.L.H.; Cerdà, A.; Imeson, A.C. Ecohydrological adaptation of soils following land abandonment in a semi-arid environment. *Ecohydrology* **2010**, *3*, 421–430. [CrossRef]

41. Osawa, T.; Kohyama, K.; Mitsuhashi, H. Trade-off relationship between modern agriculture and biodiversity: Heavy consolidation work has a long-term negative impact on plant species diversity. *Land Use Policy* **2016**, *54*, 78–84. [CrossRef]

42. Hannula, S.E.; Morriën, E.; de Hollander, M.; van der Putten, W.H.; van Veen, J.A.; de Boer, W. Shifts in rhizosphere fungal community during secondary succession following abandonment from agriculture. *ISME J.* **2017**, *11*, 2294–2304. [CrossRef]

43. Cavani, L.; Manici, L.M.; Caputo, F.; Peruzzi, E.; Ciavatta, C. Ecological restoration of a copper polluted vineyard: Long-term impact of farmland abandonment on soil bio-chemical properties and microbial communities. *J. Environ. Manag.* **2016**, *182*, 37–47. [CrossRef] [PubMed]

44. He, B.; Wang, H.; Huang, L.; Liu, J.; Chen, Z. A new indicator of ecosystem water use efficiency based on surface soil moisture retrieved from remote sensing. *Ecol. Indic.* **2017**, *75*, 10–16. [CrossRef]

45. Ramos, M.C.; Martínez-Casasnovas, J.A. Impact of land levelling on soil moisture and runoff variability in vineyards under different rainfall distributions in a Mediterranean climate and its influence on crop productivity. *J. Hydrol.* **2006**, *321*, 131–146. [CrossRef]

46. Al-Shammary, A.A.G.; Kouzani, A.Z.; Kaynak, A.; Khoo, S.Y.; Norton, M.; Gates, W. Soil Bulk Density Estimation Methods: A Review. *Pedosphere* **2018**, *28*, 581–596. [CrossRef]

47. Bienes, R.; Marques, M.J.; Sastre, B.; García-Díaz, A.; Ruiz-Colmenero, M. Eleven years after shrub revegetation in semiarid eroded soils. Influence in soil properties. *Geoderma* **2016**, *273*, 106–114. [CrossRef]

48. Jomaa, S.; Barry, D.A.; Brovelli, A.; Heng, B.C.P.; Sander, G.C.; Parlange, J.Y.; Rose, C.W. Rain splash soil erosion estimation in the presence of rock fragments. *CATENA* **2012**, *92*, 38–48. [CrossRef]

49. Poesen, J.; Van Wesemael, B.; Govers, G.; Martinez-Fernandez, J.; Desmet, P.; Vandaele, K.; Quine, T.; Degraer, G. Patterns of rock fragment cover generated by tillage erosion. *Geomorphology* **1997**, *18*, 183–197. [CrossRef]

50. Šraj, M.; Brilly, M.; Mikoš, M. Rainfall interception by two deciduous Mediterranean forests of contrasting stature in Slovenia. *Agric. For. Meteorol.* **2008**, *148*, 121–134. [CrossRef]

51. Leuning, R.; Condon, A.G.; Dunin, F.X.; Zegelin, S.; Denmead, O.T. Rainfall interception and evaporation from soil below a wheat canopy. *Agric. For. Meteorol.* **1994**, *67*, 221–238. [CrossRef]

52. Facelli, J.M.; Pickett, S.T.A. Plant Litter: Light Interception and Effects on an Old-Field Plant Community. *Ecology* **1991**, *72*, 1024–1031. [CrossRef]

53. Llorens, J.; Gil, E.; Llop, J.; Queraltó, M. Georeferenced LiDAR 3D vine plantation map generation. *Sensors* **2011**, *11*, 6237–6256. [CrossRef] [PubMed]

54. Hou, J.; Fu, B.; Liu, Y.; Lu, N.; Gao, G.; Zhou, J. Ecological and hydrological response of farmlands abandoned for different lengths of time: Evidence from the Loess Hill Slope of China. *Glob. Planet. Chang.* **2014**, *113*, 59–67. [CrossRef]

55. Otero, I.; Boada, M.; Badia, A.; Pla, E.; Vayreda, J.; Sabaté, S.; Gracia, C.A.; Peñuelas, J. Loss of water availability and stream biodiversity under land abandonment and climate change in a Mediterranean catchment (Olzinelles, NE Spain). *Land Use Policy* **2011**, *28*, 207–218. [CrossRef]

56. Beguería, S.; López-Moreno, J.I.; Lorente, A.; Seeger, M.; García-Ruiz, J.M. Assessing the effect of climate oscillations and land-use changes on streamflow in the central Spanish Pyrenees. *Ambio* **2003**, *32*, 283–286. [CrossRef] [PubMed]

57. Keesstra, S.D.; Bruijnzeel, L.A.; Huissteden, J. Meso-scale catchment sediment budgets: Combining field surveys and modeling in the Dragonja catchment, southwest Slovenia. *Earth Surf. Process. Landf.* **2009**, *34*, 1547–1561. [CrossRef]

58. Moreira, M.Z.; Sternberg, S.L.; Nepstad, D.C. Vertical patterns of soil water uptake by plants in a primary forest and an abandoned pasture in the eastern Amazon: An isotopic approach. *Plant Soil* **2000**, *222*, 95–107. [CrossRef]

59. Rambousková, H. Water dynamics of some abandoned fields of the Bohemian Karst. *Folia Geobot. Phytotax.* **1981**, *16*, 133. [CrossRef]

60. Farrick, K.K.; Price, J.S. Ericaceous shrubs on abandoned block-cut peatlands: Implications for soil water availability and Sphagnum restoration. *Ecohydrology* **2009**, *2*, 530–540. [CrossRef]

61. Ruecker, G.; Schad, T.; Alcubilla, M.N.; Ferrer, C. Natural regeneration of degraded soils and site changes on abandoned agricultural terraces in Mediterranean Spain. *Land Degrad. Dev.* **1998**, *9*, 179–188. [CrossRef]

62. Koulouri, M.; Giourga, C. Land abandonment and slope gradient as key factors of soil erosion in Mediterranean terraced lands. *CATENA* **2007**, *69*, 274–281. [CrossRef]

63. Lesschen, J.P.; Cammeraat, L.H.; Nieman, T. Erosion and terrace failure due to agricultural land abandonment in a semi-arid environment. *Earth Surf. Process. Landf.* **2008**, *33*, 1574–1584. [CrossRef]

64. Liu, Y.; Tao, Y.; Wan, K.Y.; Zhang, G.S.; Liu, D.B.; Xiong, G.Y.; Chen, F. Runoff and nutrient losses in citrus orchards on sloping land subjected to different surface mulching practices in the Danjiangkou Reservoir area of China. *Agric. Water Manag.* **2012**, *110*, 34–40. [CrossRef]

65. Arnaez, J.; Lasanta, T.; Errea, M.P.; Ortigosa, L. Land abandonment, landscape evolution, and soil erosion in a Spanish Mediterranean mountain region: The case of Camero Viejo. *Land Degrad. Dev.* **2011**, *22*, 537–550. [CrossRef]

66. Nunes, A.N.; Coelho, C.O.A.; de Almeida, A.C.; Figueiredo, A. Soil erosion and hydrological response to land abandonment in a central inland area of Portugal. *Land Degrad. Dev.* **2010**, *21*, 260–273. [CrossRef]

67. Smetanová, A.; Follain, S.; David, M.; Ciampalini, R.; Raclot, D.; Crabit, A.; Le Bissonnais, Y. Landscaping compromises for land degradation neutrality: The case of soil erosion in a Mediterranean agricultural landscape. *J. Environ. Manag.* **2019**, *235*, 282–292. [CrossRef] [PubMed]

68. Lucas-Borja, M.E.; Zema, D.A.; Carrà, B.G.; Cerdà, A.; Plaza-Alvarez, P.A.; Cózar, J.S.; Gonzalez-Romero, J.; Moya, D.; de las Heras, J. Short-term changes in infiltration between straw mulched and non-mulched soils after wildfire in Mediterranean forest ecosystems. *Ecol. Eng.* **2018**, *122*, 27–31. [CrossRef]

69. Tang, S.; Guo, J.; Li, S.; Li, J.; Xie, S.; Zhai, X.; Wang, C.; Zhang, Y.; Wang, K. Synthesis of soil carbon losses in response to conversion of grassland to agriculture land. *Soil Tillage Res.* **2019**, *185*, 29–35. [CrossRef]

70. Xie, Y.; Lin, H.; Ye, Y.; Ren, X. Changes in soil erosion in cropland in northeastern China over the past 300 years. *CATENA* **2019**, *176*, 410–418. [CrossRef]

71. García-Llamas, P.; Suárez-Seoane, S.; Taboada, A.; Fernández-Manso, A.; Quintano, C.; Fernández-García, V.; Fernández-Guisuraga, J.M.; Marcos, E.; Calvo, L. Environmental drivers of fire severity in extreme fire events that affect Mediterranean pine forest ecosystems. *For. Ecol. Manag.* **2019**, *433*, 24–32. [CrossRef]

72. Trukhachev, A. Methodology for Evaluating the Rural Tourism Potentials: A Tool to Ensure Sustainable Development of Rural Settlements. *Sustainability* **2015**, *7*, 3052–3070. [CrossRef]

73. Grimm, N.B.; Faeth, S.H.; Golubiewski, N.E.; Redman, C.L.; Wu, J.; Bai, X.; Briggs, J.M. Global Change and the Ecology of Cities. *Science* **2008**, *319*, 756–760. [CrossRef]

74. Ackermann, O.; Zhevelev, H.M.; Svoray, T. Agricultural systems and terrace pattern distribution and preservation along climatic gradient: From sub-humid mediterranean to arid conditions. *Quat. Int.* **2019**, *502*, 319–326. [CrossRef]

75. Ackermann, O.; Maeir, A.M.; Frumin, S.S.; Svoray, T.; Weiss, E.; Zhevelev, H.M.; Horwitz, L.K. The Paleo-Anthropocene and the Genesis of the Current Landscape of Israel. *J. Landsc. Ecol.* **2017**, *10*, 109–140. [CrossRef]

76. Ackermann, O.; Zhevelev, H.M.; Svoray, T. Sarcopoterium spinosum from mosaic structure to matrix structure: Impact of calcrete (Nari) on vegetation in a Mediterranean semi-arid landscape. *CATENA* **2013**, *101*, 79–91. [CrossRef]

77. Cerdà, A. The effect of patchy distribution of *Stipa tenacissima* L. on runoff and erosion. *J. Arid Environ.* **1997**, *36*, 37–51. [CrossRef]

78. Ibáñez, J.; Contador, J.F.L.; Schnabel, S.; Fernández, M.P.; Valderrama, J.M. A model-based integrated assessment of land degradation by water erosion in a valuable Spanish rangeland. *Environ. Model. Softw.* **2014**, *55*, 201–213. [CrossRef]
79. Di Prima, S.; Rodrigo-comino, J.; Novara, A.; Iovino, M.; Pirastru, M.; Keesstra, S.; Cerdà, A. Soil Physical Quality of Citrus Orchards Under Tillage, Herbicide, and Organic Managements. *Pedosphere* **2018**, *28*, 463–477. [CrossRef]
80. Kasirajan, S.; Ngouajio, M. Polyethylene and biodegradable mulches for agricultural applications: A review. *Agron. Sustain. Dev.* **2012**, *32*, 501–529. [CrossRef]

Article

Runoff and Solute Outputs under Different Land Uses: Long-Term Results from a Mediterranean Mountain Experimental Station

Estela Nadal-Romero *, Makki Khorchani, Teodoro Lasanta and José M. García-Ruiz

Instituto Pirenaico de Ecología, Procesos Geoambientales y Cambio Global, IPE-CSIC, 50059 Zaragoza, Spain; makki.khorchani@ipe.csic.es (M.K.); fm@ipe.csic.es (T.L.); humberto@ipe.csic.es (J.M.G.-R.)
* Correspondence: estelanr@ipe.csic.es; Tel.: +34-976-369-393

Received: 1 April 2019; Accepted: 5 May 2019; Published: 9 May 2019

Abstract: Water availability and quality in Mediterranean environments are largely related to the spatial organisation of land uses in mountain areas, where most water resources are generated. However, there is scant data available on the potential effects of land use changes on surface water chemistry in the Mediterranean mountain region. In order to address this gap in the research, this study investigates the effects of various mountain Mediterranean land covers/land uses on runoff water yielded and water chemistry (solute concentrations and loads) using data from the Aísa Valley Experimental Station (Central Pyrenees) for a long-term period (1991–2011). Nine land covers have been reproduced in closed plots, including dense shrub cover, grazing meadows, cereal, fallow land, abandoned field, shifting agriculture (active and abandoned) and 2 burned plots (one burned in 1991 and the second one burned twice in 1993 and 2001). Results show that all solute concentrations differed among land uses, with agricultural activity producing significantly higher solute loads and concentrations than the other types. Two groups have been identified: (i) the lowest solute concentrations and the smallest quantities of solute loads are recorded in the dense shrub cover, the plot burned once (at present well colonized with shrubs), meadows and abandoned field plots; (ii) the plot burned twice registered moderate values and the highest solute concentrations and loads are found in cereal, fallow land and shifting agriculture plots. Water chemistry is clearly dominated by Ca^{2+} and $HCO_3{}^-$ concentrations, whereas other solutes are exported in very low quantities due to the poor nutrient content of the soil. These results complete the information published previously on soil erosion under different land uses in this experimental station and help to explain the evolution of land cover as a consequence of shifting agriculture, cereal farming on steep slopes and the use of recurrent fires to favour seasonal grazing. They also suggest that promoting the development of grazing and cutting meadows is a good strategy to reduce not only soil erosion but also the loss of nutrients.

Keywords: solutes; land uses; soil erosion; water quality; Central Pyrenees

---

1. Introduction

The quality and availability of water resources constitute one the most important environmental problems in many countries of the world (i.e., [1]). In the Mediterranean basin, water availability is scarce, with strong seasonal contrasts and most of the flows are produced in mountain rivers [2,3], while water demand is constantly increasing, especially in the lowlands, due to the growth in population and urbanization, the expansion of irrigated areas and greater water consumption of the industrial and tourism sectors [4–6]. García-Ruiz et al. [7] indicated that Mediterranean countries are among the most under threat worldwide for water stress, due to high inter-annual and seasonal rainfall variability, revegetation processes and a predicted decrease in river flows in the coming decades.

Giorgi [8] considered, on a global scale, the Mediterranean basin as a "hot spot" due to the limitation of water resources.

Furthermore, it is well known, that the evolution of surface water resources is controlled by two groups of environmental factors that are variable on a temporal scale: climate and land uses and land covers (hereafter, LULC). In the Mediterranean basin, future scenarios suggest that water resources will decrease as temperatures and evapotranspiration increase, while snow cover decreases [7,9,10]. On the other hand, LULC show an increase in vegetation cover due to the advance of shrubs and forests in abandoned lands [11,12]. The increase in vegetation cover implies a greater evapotranspiration and interception [13,14] and higher water consumption by vegetation [15], thus reducing the surface flows of Mediterranean rivers and the water for filling reservoirs [16].

Revegetation processes in European Mediterranean mountain areas constitute an important problem for land managers, not only for their implications in the decrease of water resources [15–17] but also for the increase in fires [18], landscape homogenization, with less biodiversity and aesthetic quality [19,20], degradation and loss of grazing resources [21,22] and ecosystem services [23,24]. To reduce the negative effects of revegetation processes, land managers have implemented public policies that try to control the revegetation process: prescribed fires [22,25] and shrub clearings [26–28].

On the other hand, in Southern and Eastern Mediterranean countries, agriculture is still practiced in many mountain areas to supply local population or for export, cultivating marginal areas (with low soil fertility and steep slopes) and in extreme cases, even using shifting agriculture systems. However, in other mountains, a process of land abandonment has started and is expected to intensify in the coming decades [29,30]. Most of the research related to land abandonment and revegetation processes in Mediterranean areas has focused on changes in the soil properties [31,32] and biodiversity and vegetation cover [33] and much less work has been done on soil erosion over the long-term [17,34]. Despite being generally accepted that solute outputs can be as important as other forms of surface erosion [35], studies that address the potential effects of land use changes on surface water chemistry and quality have been carried out to a lesser extent in the Mediterranean mountain region.

The loss of nutrients from different land uses (cultivated or uncultivated soils) contributes to a decrease in soil productivity and encourages soil erosion processes and other signs of ecosystem deterioration [36]. Many authors confirmed the extreme importance of LULC management in explaining the loss of nutrients and water quality (i.e., [37]) and changes in water chemistry associated with land use conversion have been demonstrated worldwide [38–40]. Most of these studies were carried out at catchment scale [41–43] and very little data are available at plot scale (which provides better knowledge on the effects from different land uses and a comparison with the same environmental conditions). For example, Oliveira et al. [44] and Ferreira et al. [45] observed an increase in solute loads due to anthropogenic-related land covers (agricultural and urban land). Likewise, Merchán et al. [46] showed that agricultural land use tends to increase the concentration of both solutes and suspended sediment. Nevertheless, the study of solute sources encounters many difficulties in anthropogenically-disturbed areas, in great part due to the wide range of LULC at catchment scale and the difficulty of directly observing the origin and pathways of solutes. One possible strategy is to reproduce different land uses in experimental plots and to measure runoff and sediment outputs during each storm event, while soil and gradient remains fixed. We are aware that the use of plots poses significant methodological problems and that an upscaling approach would be needed to interpret solute and suspended sediment values and sources at a catchment scale.

One of the great challenges for the future is to ensure sufficient water resources (quantity and quality), through the sustainable management of mountain areas [47]. To address this research gap, the present study aimed at providing further insights into solute concentration and total solute loads in nine different land uses and land covers (representing traditional and present LULC) in a Mediterranean mountain area (dense shrub cover, grazing meadows, cereal, fallow land, abandoned field, shifting agriculture (active and abandoned) and two burned plots). We hypothesized that different land uses conditioned water chemistry, solute concentration and export rates, with traditional agricultural uses recording the highest solute concentrations and loads. To improve our knowledge of the influence of land uses and land management on water chemistry, we analysed the information obtained from overland flow (surface runoff) in the Aísa Valley Experimental Station for 20 years (1991–2011), allowing us to define the solute patterns in both the traditional and present management systems.

2. Materials and Methods

2.1. The Aísa Valley Experimental Station (1991–2011)

The Aísa Valley Experimental Station (hereafter AVES) was installed in a representative slope located on a field abandoned 30 years ago. It is located at 1149 m a.s.l. on a south-facing slope completely covered by a dense shrubland of *Genista scorpius* L. and *Rosa gr. canina* L. The parent material is Eocene Flysch (alternate thin layers of sandstones and marls), at a 30% slope gradient. The climate is sub-Mediterranean with mountain characteristics. The average annual precipitation is 1100 mm and the average annual temperature is 10 °C (for more details see [17]).

Nine closed plots (3 m width × 10 m length, with the same slope morphology, soil type and topographic conditions) were installed, including at the lower end a Gerlach trap and a system of tipping buckets connected to data loggers in order to record the runoff of each plot continuously. A pluviometer was also connected to a data logger. In the lowest part, a plastic collector to store runoff and sediment during rainfall events was installed in each plot. After each rainfall, a representative sample (about 2 L) was collected and analysed in the laboratory for suspended sediment concentration, total solute concentration and solute composition (mean number of runoff events in the study period equals to 354 (per plot) and mean number of runoff events per year and plot equals to 18 ± 8). Chemical water analysis for a set of relevant elements was carried out (Na^+, K^+, Ca^{2+}, Mg^{2+}, HCO_3^-, Cl, SO_4^{2-}, NO_3^-, SiO_2, $N\text{-}NO_3^-$, $N\text{-}NO_2^-$, $N\text{-}NH_4^-$ and PO_4^{3-}) over twenty years by means of ion chromatography (Metrohm 861 Advanced Compact IC) at the Laboratory of the Instituto Pirenaico de Ecología (IPE-CSIC).

Plots reproduced nine LULC (Figure 1) from traditional and present management systems in Mediterranean areas: (i) dense shrub cover (DS) (with the unaltered and original vegetation growing after land abandonment decades ago), (ii) grazing meadows (GM) (a regularly grazed field with manual annual harvest), (iii) cereal plot (C) (adding chemical fertilizers and alternating with the fallow land plot), (iv) fallow land (FL) (alternating with the cereal plot), (v) abandoned field (AF) (it was cultivated in the first year and then abandoned to initiate natural revegetation), (vi) shifting agriculture (SA) (traditionally only fertilized with ashes), (vii) abandoned shifting agriculture (ASA) (a cultivated shifting field that was abandoned after 4 years of cultivation to initiate natural revegetation), (viii) burned plot 1 (B1) (burned in 1991) and (ix) burned plot 2 (B2) (burned twice in 1993 and 2001) (Figure 2). Both burned plots were affected by plant recolonization following the fire.

In the cereal plot, chemical fertilizer was added annually with $N\text{-}NH_4$ (30 kg ha^{-1}), P_2O_5 (60 kg ha^{-1}) and K_2O (20 kg ha^{-1}) as the main constituents, with very small quantities of Mg^{2+}, S and Fe. Given that the fallow land plot alternated yearly with the cereal plot, its soil was also enriched with chemical fertilizer each two years. The abandoned plot (previously cultivated with cereal) received similar quantities of chemical fertilizer as the cereal plot during the years it was cultivated.

In addition, 5 top soil samples (1991, non-disturbed due to experimentation) were taken spatially distributed in the slope where the AVES was located and physico-chemical analyses were made, including texture, pH, organic matter and nitrogen, $CaCO_3$ and P, K, Mg content.

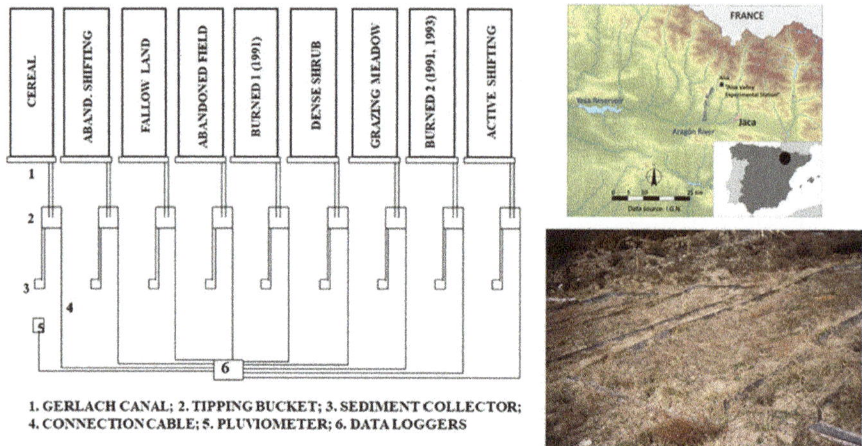

Figure 1. Study area and Aísa Valley Experimental Station scheme and overview of the station.

Figure 2. Plant cover evolution in the different land uses in two different stages.

2.2. Statistical Analysis

Data from each plot were analysed statistically to provide annual averages (mean, median and standard deviation). Likewise, box plots were used to provide a representation of the annual average solute concentration data variability.

As the assumption of normal distribution could not be met for most variables, as checked by a Shapiro-Wilk normality test, the non-parametric Kruskal-Wallis one-way analysis was used to identify which LULC differ statistically from each other. Correlations between parameters were checked and tested by the Spearman's rank correlation test. In all cases, differences were taken to be statistically significant at $p < 0.05$. Principal component analysis (PCA) was also performed to determine first correlations among the measured variables and to elucidate major variation patterns in terms of land uses.

3. Results

3.1. Soil Properties and Land Cover Evolution

Physical and chemical soil properties were analysed in 1991 for the original hillslope where the experimental station was located (revegetated area after some decades of land abandonment) (Table 1). Grain size distributions showed a predominance of sand fraction (loam texture): 40.5% was sand, 34.4% was silt and 25.1% was clay. Chemical soil analysis demonstrated that the soils in the study area had very poor contents of P, K and Mg, whilst they were abundant in N owing to the presence of leguminous plants (*G. scorpius*). Organic matter content was 2.3%. Conversely, the soils showed a very high carbonate content (32.9%) corresponding to soils developed on a calcareous substratum.

Table 1. Mean and standard deviation values of the studied soil parameters in 1991.

Soil Parameters (1991)	Mean and Standard Deviation
pH	7.9 ± 0.3
% $CaCO_3$	32.9 ± 6.2
% N	1.8 ± 0.7
% Organic matter	2.3 ± 1.1
C/N ratio	7.1 ± 0.7
P (ppm)	4.4 ± 3.7
K (ppm)	1.3 ± 0.5
Mg (ppm)	0.8 ± 0.2
% Sand	40.5 ± 2.7
% Silt	34.4 ± 4.9
% Clay	25.1 ± 5.9

Figure 2 presents the evolution of plant cover in the different plots during the study period, with some remarkable results: (i) surprisingly, the dense shrub cover plot gradually evolved with open spaces due to the senescence of the shrub (together with a higher herbaceous cover) (Figure 2). In addition, partial replacement of *G. scorpius* by *Juniperus communis* was observed; (ii) the abandoned cereal plot was rapidly revegetated (first by a dense herbaceous cover and later with shrubs) and tended to be similar to that of the dense shrub cover at the end of the study period thanks to the fertilizer added when it was cultivated (Figure 2); (iii) the abandoned shifting agriculture evolved much slower than the abandoned cereal plot due to the differences observed in soil properties and fertilizers, confirming soil deterioration in slopes cultivated under slash-and-burn practices; this plot showed a rapid herbaceous recovery, although shrubs found severe problems in colonizing the plot; and (iv) in the burned plots a rapid recovery of vegetation was observed in the first few years after the fire. Interestingly, during the study period, the vegetation cover was lower in the plot burned twice (1993 and 2001) than in the plot that was burned only once in 1991 (Figure 2). In any case, the burned plot 2 also shows a positive trend in vegetation recovery after each fire. As expected, plant cover in the fallow land, meadow, cereal and shifting agriculture plots did not change throughout the study period and only changes due to vegetative cycle of the crop were observed (Figure 2).

3.2. Runoff and Suspended Sediment Concentration Updating

Mean annual rainfall value for the study period was 1181 mm. Data on runoff response for the period 1996–2009 and for only 7 plots has already been presented in Nadal-Romero et al. [17]. Table 2 presents an update of these values with data relating to the full period (1991–2011 with small variations) and 9 LULC. Concerning runoff water yielded, three groups of plots were differentiated: (i) the lowest runoff values were recorded in the dense shrub cover (63.5 L m^{-2}), burned 1 (burned in 1991, with dense vegetation cover) (68.4 L m^{-2}) and grazing meadow (89.2 L m^{-2}) plots, (ii) moderate values were recorded in the abandoned field (137.1 L m^{-2}), abandoned shifting agriculture (146.2 L m^{-2}) and burned 2 (158.3 L m^{-2}) plots; and (iii) the third group includes the plots related to agricultural activities showing the highest values: fallow land (181.0 L m^{-2}), cereal (237.5 L m^{-2}) and active shifting agriculture (239.0 L m^{-2}) plots. Table 2 also refers to suspended sediment and solute concentration, as well as to the total sediment yields and the percentage of solutes to total sediment production. Suspended sediment concentration (mg L^{-1}) tends to behave in a similar way to runoff, although some differences must be noted: (i) in the first group, with low values of suspended sediment concentration, the abandoned field plot (37.8 mg L^{-1}) adds to the dense shrub cover (26.5 mg L^{-1}), the burned 1 (28.3 mg L^{-1}) and the grazing meadow (35.2 mg L^{-1}) plots; (ii) in the second group, the cereal plot (138.6 mg L^{-1}) recorded similar values to that of abandoned shifting agriculture (118.2 mg L^{-1}); and (iii) the third group comprises burned 2 (151.5 mg L^{-1}), the active shifting agriculture (223.1 mg L^{-1}) and the fallow land (298.6 mg L^{-1}) plots. Information on the standard deviation for the suspended sediment concentration reveals an extremely high variability, particularly in the case of the plots with greatest erosion (fallow land and shifting agriculture plots followed by the burned plot 2).

3.3. Solute Water Chemistry and Total Sediment Production

Data on solute concentrations can be found in Table 2, with significant differences among LULC. In this case, the statistical analyses indicated that only two groups of land uses can be distinguished with small differences in respect of those identified for water and annual sediment production:

(i)　The lowest solute concentration values were recorded in the grazing meadow (115.6 mg L^{-1}) and dense shrub cover (118.1 mg L^{-1}) plots, where agricultural activity was absent during all the study period and a dense plant cover was recorded. Mid-range values were obtained in the abandoned cereal field (133.1 mg L^{-1}) and the two burned plots (1 and 2) (130.6 and 136.2 mg L^{-1}, respectively).

(ii)　The highest values were recorded in the cereal plot (192.6 mg L^{-1}), fallow (196.1 mg L^{-1}) and shifting agriculture (active and abandoned) (159.7 and 181.9 mg L^{-1} respectively), where chemical fertilizer or ash were added.

The total sediment yield for the nine plots (g m^{-2}) appear in Table 2, providing a general perspective of the influence of different LULC on sediment losses. Two groups of plots were also distinguishable: (i) the dense shrub cover plot produced the lowest mean annual sediment yield (9.8 g m^{-2}), followed by burned 1 (13.3 g m^{-2}), the grazing meadow plot (16.2 g m^{-2}) and the abandoned field (27.2 g m^{-2}). (ii) Intermediate yields (already two-three fold higher than the previous values) were found for the plot burned twice (64.5 g m^{-2}) and the cereal plot (81.6 g m^{-2}), followed by the highest yields that were found for the fallow land (100.9 g m^{-2}), the abandoned shifting agriculture (102.7 g m^{-2}) and the active shifting agriculture (143.8 g m^{-2}) plots. The proportion of solutes in relation to the total sediment output varied significantly, showing that plots with low values of sediment exportation have a greater loss of solutes than those of suspended sediment: dense shrub cover, grazing meadow, abandoned field and burned plot 1, where solutes represent more than 63% of the total sediment outputs.

Table 2. Mean values and standard deviations of runoff, suspended sediment concentration, solute concentration, total sediment and % of solutes for the period of 1991–2011 (except for abandoned field whose data are from 1992–2011, burned 2 whose data are from 1993–2011 and abandoned shifting agriculture whose data are from 1995–2011). Means with the different lower-case letter within a column are significantly different at the 0.05 level of significance ($p < 0.05$).

	Annual Runoff ($L\ m^{-2}$)	Suspended Sediment Concentration ($mg\ L^{-1}$)	Solute Concentration ($mg\ L^{-1}$)	Total Annual Sediment ($g\ m^{-2}$)	% Solutes
Dense shrub cover	63.5 ± 32.0 (a)	26.5 ± 24.1 (a)	118.1 ± 27.9 (a)	9.8 ± 5.63 (a)	72.1 (a)
Burned 1 (1991)	68.4 ± 28.9 (a)	28.3 ± 39.5 (a)	130.6 ± 27.5 (a)	13.3 ± 6.08 (a)	70.1 (a)
Grazing meadow	89.2 ± 47.4 (a)	35.2 ± 41.3 (a)	115.6 ± 33.8 (a)	16.2 ± 7.55 (a)	63.4 (a)
Abandoned field	137.1 ± 62.0 (b)	37.8 ± 37.6 (a)	133.1 ± 33.3 (a)	27.2 ± 13.31 (a)	63.0 (a)
Burned 2 (1993–2001)	158.3 ± 88.4 (bc)	151.5 ± 221.0 (bc)	136.2 ± 43.0 (a)	64.5 ± 46.37 (b)	31.3 (b)
Cereal	237.5 ± 205.9 (c)	138.6 ± 118.5 (b)	192.6 ± 52.0 (b)	81.6 ± 51.54 (b)	40.0 (b)
Fallow land	181.0 ± 109.9 (bc)	298.6 ± 398.4 (c)	196.1 ± 41.1 (b)	100.9 ± 79.26 (b)	29.2 (b)
Abandoned shifting agriculture	146.2 ± 100.9 (bc)	118.2 ± 93.3 (b)	159.7 ± 24.7 (b)	102.7 ± 114.5 (b)	22.4 (b)
Active shifting agriculture	239.0 ± 143.8 (c)	223.1 ± 298.9 (bc)	181.9 ± 26.6 (b)	143.8 ± 122.7 (b)	29.1 (b)

Table 3. Spearman Correlations between pH and electrical conductivity (EC) and a set of relevant elements of water ($p < 0.05$ *; $p < 0.01$ **).

	pH	CE	HCO_3^-	Cl	SO_4^{2-}	Ca^{2+}	Mg^{2+}	Na^+	K^+	PO_4^{3-}	$N-NO_3^-$	$N-NO_2^-$	$N-NH_4^+$	SiO_2
pH														
CE	0.169 *													
HCO_3^-	0.464 **	0.641 **												
Cl	−0.280 **	0.418 **	-0.193 **											
SO_4^{2-}	−0.366 **	0.357 **	0.445 **	0.008										
Ca^{2+}	0.047	0.675 **	0.514 **	0.131	0.266 *									
Mg^{2+}	0.085	0.686 **	0.457 **	0.354 **	0.330 **	0.544 **								
Na^+	0.050	0.589 **	0.343 **	0.325 **	0.542 **	0.454 **	0.464 **							
K^+	−0.194 **	0.121	0.002	0.247 **	0.116	0.028	0.134	0.179 *						
PO_4^{3-}	0.030	0.104	0.077	0.021	0.092	0.081	0.027	0.009	0.115					
$N-NO_3^-$	0.188 *	−0.035	0.076	−0.003	0.302 **	−0.0115	−0.036	−0.044	0.166 *	0.329 **				
$N-NO_2^-$	0.118	0.072	0.137	0.035	0.213 **	0.027	0.039	0.072	0.148 *	−0.066	0.402 **			
$N-NH_4^-$	−0.072	0.013	−0.024	−0.004	−0.040	0.039	−0.024	−0.100	−0.009	−0.018	0.115	0.521 **		
SiO_2	−0.069	−0.278 **	−0.214 **	−0.349 **	−0.325 **	0.104	−0.219	−0.289 **	−0.042	0.195 **	−0.205 **	−0.270 **	−0.048	

Finally, Table 2 also shows that there is more variability (higher standard deviations) in the case of suspended sediment concentrations than in solutes.

Table 3 presents the relationship among chemical water variables and solute concentrations. Concentrations of cations were, in general, positively related to each other. Negative correlations were found between HCO_3^- and Cl^- concentrations and SIO_2 and electrical conductivity, HCO_3^-, Cl^-, SO_4^{2-}, Na^+, PO_4^{3-}, $N\text{-}NO_3^-$ and $N\text{-}NO_2^-$. No statistical correlations were found between Cl^- and SO_4^{2-} concentrations. In addition, significant correlation between $N\text{-}NO_3^-$ and SO_4^{2-} was observed.

Figure 3 presents box plots to summarize water quality responses at the nine LULC (solute concentration, mg L^{-1}) and Table 4 informs on the total losses of the most representative solutes for each LULC (g m^{-2}). The most significant solute outputs were in the form of HCO_3^-, with losses close to 20 g m^{-2} from cereal, shifting agriculture and fallow land. Obviously Ca^{2+} was also carried out in high quantities (values between 5–8.5 g m^{-2} in the same plots cited above), while the rest of solutes showed lower values, especially Na^+ and Mg^{2+} (due to the low content in the soil). Figure 4 illustrated the $Ca^{2+}\text{-}HCO_3^-$ dominance of the solute losses. It is noteworthy that in the LULC generally affected by low solute outputs (dense shrub cover, grazing meadow, abandoned field and abandoned shifting agricultural plots) HCO_3^- represents more than 90% of total anions exported (see Figure 4).

Annual outputs of each of the solutes were significantly higher from the cereal, fallow land and shifting agriculture plots than from the other LULC (Table 4). For example, the production of HCO_3^-, was 3-fold higher in traditional and current agricultural uses than in dense shrub cover and grazing meadows. Similar differences were recorded in most of the solute export rates. Annual yield from vegetated plots (dense shrub cover and grazing meadow) was very low.

The relative contributions of dissolved inorganic nitrogen forms (Figure 5) differed among the LULC, with $N\text{-}NO_3^-$ being predominant in all land uses (concentrations in $N\text{-}NO_3^-$ were one order of magnitude higher than in the other nitrogen forms). Lower values were measured in the shifting agriculture and vegetated plots and the average yield was double in cereal and fallow land than in the other land uses. The most important result is that outputs of inorganic nitrogen forms are higher in the plots with active agriculture. Moreover, $N\text{-}NH_4^+$ is mainly yielded from the abandoned shifting agriculture and the burned plot 2s. The cereal and fallow land plots, as well as the burned plot 2, also lost relatively high quantities of organic phosphorous, particularly compared with the dense shrub cover, the grazing meadow and the active shifting agriculture plots.

To confirm and summarize all this information, a principal component analysis (PCA) was carried out with all the values related to solute concentrations (runoff, pH, CE, cations, anions and nitrogen forms). Figure 6 shows a plot of the eigenvector in the plane of the first two components together with the PC scores in the plane of PC1 and PC2. On the first component, which explained 21.686% of the variance, CE, runoff and all the cations and anions are positive. The second component explained 12.846% of the variance: it has large positive eigenvector values mainly for the main nitrogen forms and pH. The different LULC were successfully distinguished in the plane of PC1 and PC2: (i) dense shrub cover and grazing meadow plots (both located in the negative side of the components) and moderate plots: the two burned plots and the abandoned field plot; and (ii) conversely, the cereal and fallow plots were located in the positive side of the components, close to the active and abandoned shifting agriculture plots (indicating high values).

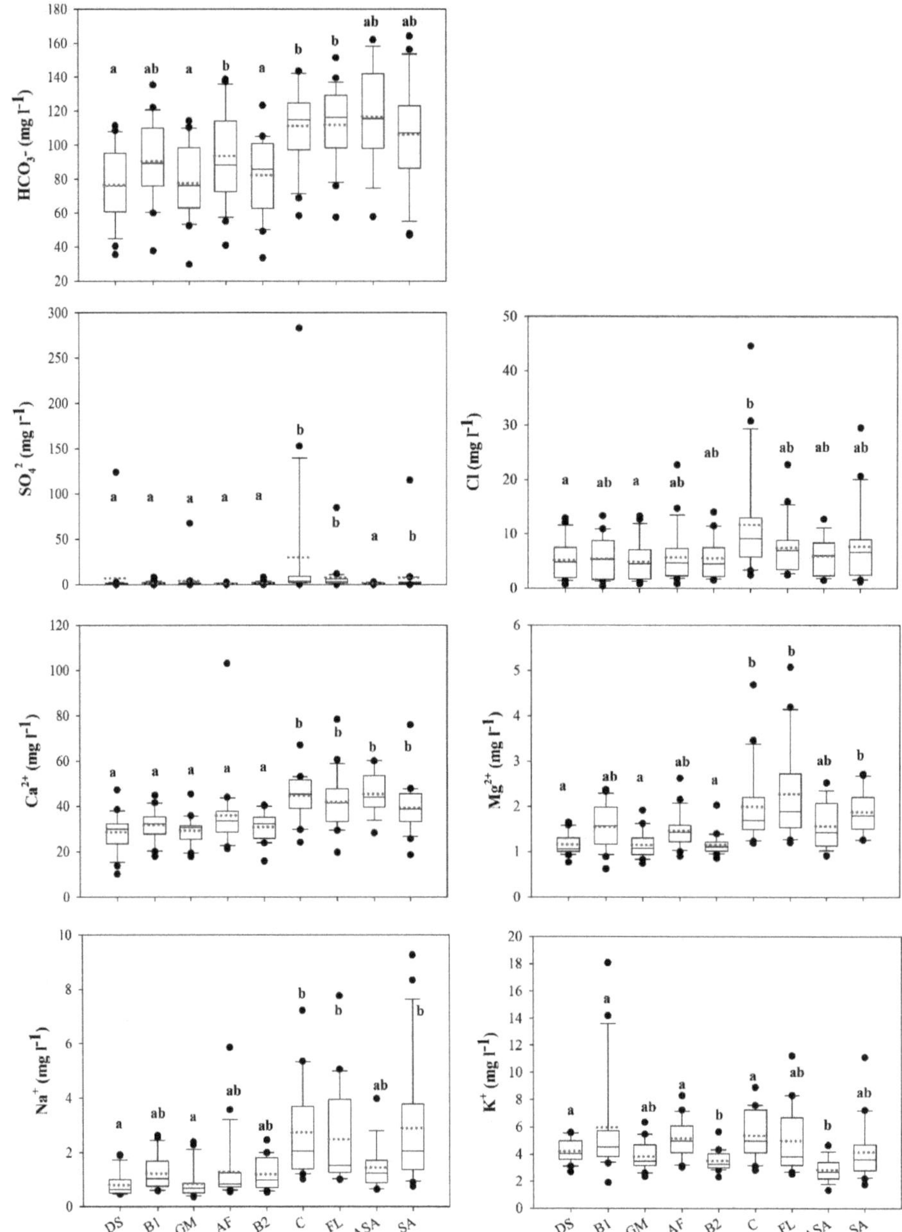

Figure 3. Solute concentrations (mg L^{-1}) from various land uses for the period 1991–2011 (red dotted line: mean value; black line: median values; black points: outliers; borders of the box are 25 and 75 percentiles; and borders of the box plot are the 10 and 90 percentiles). DS: Dense shrub cover; B1: burned plot 1; GM: grazing meadows; AF: abandoned field; B2: burned plot 2; C: Cereal; FL: fallow land; ASA: abandoned shifting agriculture; AS: active shifting. Means with the different lower-case letter within are significantly different at the 0.05 level of significance ($p < 0.05$).

Table 4. Mean and standard deviations of solute losses (g m^{-2}) for the period 1991–2011 (except for abandoned field whose data are from 1992–2011, burned 2 whose data are from 1993–2011 and abandoned shifting agriculture whose data are from 1995–2011). Means with the different lower-case letter within a column are significantly different at the 0.05 level of significance ($p < 0.05$).

	HCO_3^-	Cl^-	SO_4^{2-}	Ca^{2+}	Mg^{2+}	Na^+	K^+
Dense shrub cover	4.6 ± 2.7 (a)	0.3 ± 0.2 (a)	0.1 ± 0.2 (a)	1.7 ± 1 (a)	0.1 ± 0.04 (a)	0.04 ± 0.03 (a)	0.3 ± 0.2 (a)
Burned 1 (1991)	6.3 ± 3.0 (ab)	0.1 ± 0.3 (ab)	0.2 ± 2.0 (a)	2.2 ± 1.0 (a)	0.1 ± 0.1 (ab)	0.1 ± 0.1 (ab)	0.4 ± 0.3 (a)
Grazing meadow	6.6 ± 3.0 (a)	0.4 ± 0.4 (a)	0.2 ± 0.8 (a)	2.6 ± 1.1 (a)	0.1 ± 0.1(a)	0.07 ± 0.03 (a)	0.3 ± 0.2 (ab)
Abandoned field	12.3 ± 6.6 (ab)	0.7 ± 0.6 (ab)	0.1 ± 0.1 (a)	5.0 ± 4.0 (a)	0.2 ± 0.1 (ab)	0.1 ± 0.1 (ab)	0.7 ± 0.4 (a)
Burned 2 (1993–2001)	14.2 ± 9.9 (a)	0.9 ± 0.8 (ab)	0.1 ± 0.8 (a)	5.2 ± 3.2 (a)	0.2 ± 0.1 (a)	0.2 ± 0.2 (ab)	0.6 ± 0.3 (b)
Cereal	19.2 ± 9.5 (b)	1.8 ± 1.1 (b)	2.3 ± 5.2 (b)	7.8 ± 3.3 (b)	0.4 ± 0.3 (b)	0.5 ± 0.3 (b)	0.9 ± 0.5 (a)
Fallow land	17.1 ± 10.5 (b)	1.3 ± 1.0 (ab)	0.8 ± 1.8 (b)	6.5 ± 3.6 (b)	0.4 ± 0.3 (b)	0.5 ± 0.5 (b)	0.7 ± 0.5 (ab)
Abandoned shifting agriculture	14.2 ± 7.6 (ab)	0.7 ± 0.6 (ab)	0.1 ± 0.2 (a)	5.6 ± 3.5 (b)	0.3 ± 0.2 (ab)	0.2 ± 0.1 (ab)	0.4 ± 0.3 (b)
Active shifting agriculture	23.7 ± 16.0 (ab)	2.0 ± 2.0 (ab)	0.5 ± 0.8 (ab)	8.5 ± 5.2 (b)	0.4 ± 0.2 (b)	0.7 ± 0.8 (b)	1.0 ± 0.8 (ab)

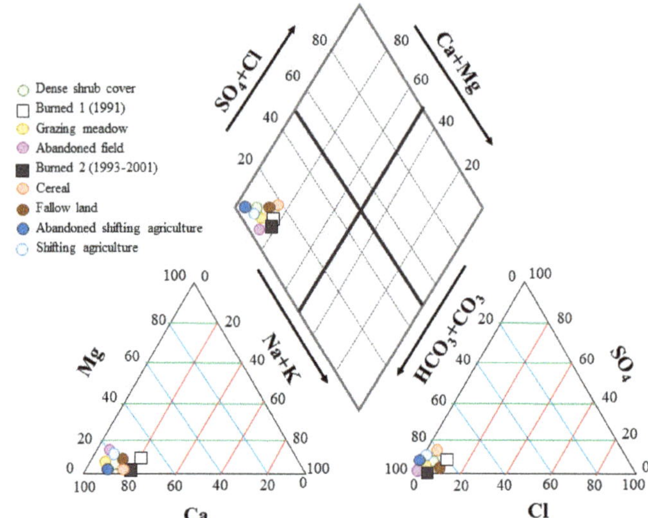

Figure 4. Piper diagram with mean solute losses from various land uses for the period 1991-2011.

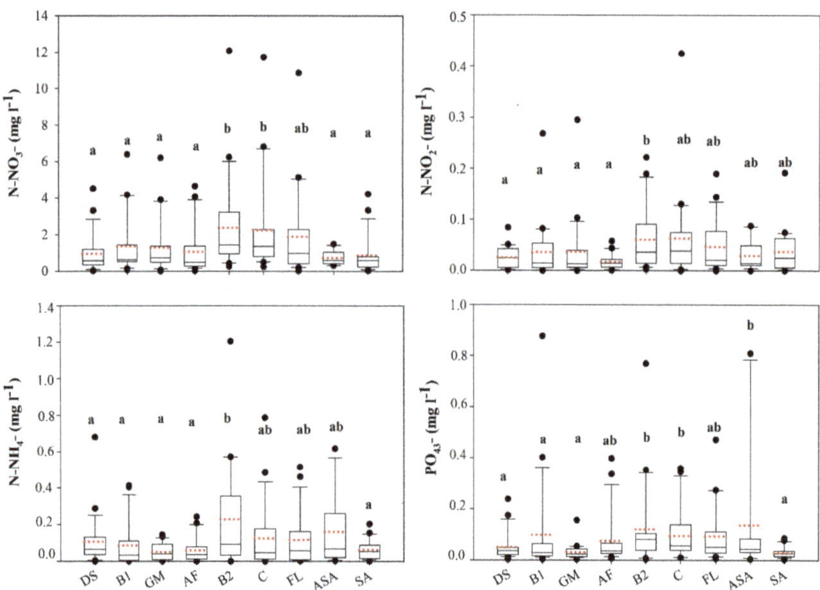

Figure 5. Nitrogen and phosphorus concentrations (mg L^{-1}) from various land uses for the period 1991–2011 (red dotted line: mean value; black line: median values; black points: outliers; borders of the box are 25 and 75 percentiles; and borders of the box plot are the 10 and 90 percentiles). DS: Dense shrub cover; B1: burned plot 1; GM: grazing meadows; AF: abandoned field; B2: burned plot 2; C: Cereal; FL: fallow land; ASA: abandoned shifting agriculture; AS: active shifting. Means with the different lower-case letter within are significantly different at the 0.05 level of significance ($p < 0.05$).

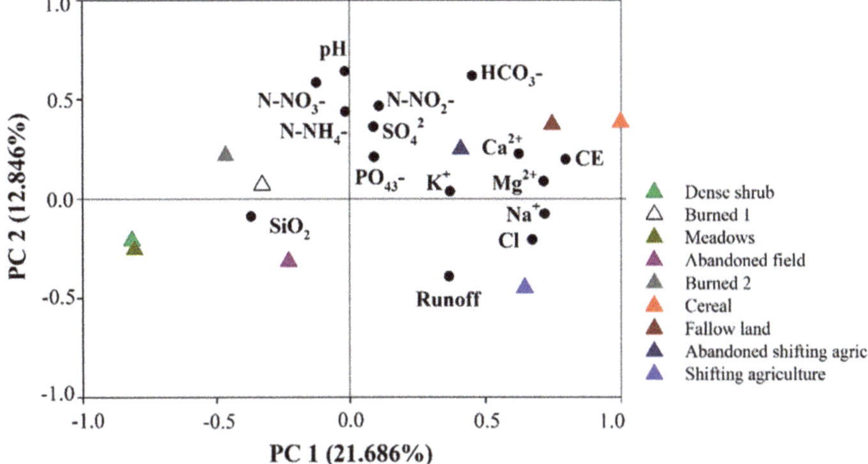

Figure 6. Eigenvectors from the principal component analysis (PCA) and PC scores plotted in the plane of the first two components.

4. Discussion

This study confirms the hypothesis that different LULC conditioned runoff erosion and solute concentration and yield. Nevertheless, it should be highlighted that the use of experimental plots for estimating runoff, suspended sediment and solutes has many limitations [48]. In our study, the characteristics of the plots, especially the size ($30\ m^2$), limit the validity of the data (since within-ground water cannot be included). A different limitation is related to the sediment exhaustion due to the system of closed plots. Although this fact has not been analysed in the study, Nadal-Romero et al. [17] observed a negative trend in sediment yield in some plots suggesting that sediment exhaustion occurred several years after initiation of the experiment. This negative trend in some cases could be related to vegetation recovery (followed by a stabilization) but in other cases (as cereal and fallow land) a clear decrease is observed during the first years of study, suggesting sediment exhaustion related to the specific conditions of the station. For this reason, the authors considered to terminate the experiment avoiding the analysis of the last two years on the database. Nevertheless, the plots enable different LULC to be studied individually and the results are good indicators of the relative differences among these LULC for comparative purposes. Likewise, our results are consistent with other studies in the Mediterranean region carried out at catchment scale [41,46].

Previous studies in this experimental station already confirmed that LULC controls runoff and suspended sediment production [17]. In the case of solute concentrations and loads, the results obtained in this study demonstrate that there are highly significant differences between the LULC with little soil and plant cover disturbances (i.e., dense shrub cover or grazing meadows) and those greatly disturbed by human activities (i.e., cereal or shifting agriculture). In the AVES, the agriculture plots (cereal, fallow and shifting agriculture) produced significantly higher solute concentrations than vegetated areas. These results are in line with previously published studies [37,39,45]. Risking et al. [49] indicated that intensive cropland agriculture commonly increases water solute concentrations and export rates. Other authors have also reported higher concentrations in agricultural than in non-agricultural areas under similar geological and climatological conditions [46]. Indeed, agriculture is a major source of contamination for water bodies worldwide (i.e., [50]) due to the use of nutrients (most studies only focus on N and P), agrochemicals and the agricultural activity, increasing runoff production, erosion rates and the export of solutes and changing water chemistry. Moreover, the removal (ploughing) of the soil year by year in the agricultural plots causes the renewal of surface soil characteristics and, as a consequence, new nutrients are available to be carried out of the plot.

It is also interesting to note the solute behaviour observed in the burned plots. Numerous studies evaluating the effects of fire on solute losses have been published in the last decades (e.g., [51–53]), all of which confirmed that solute losses were much higher on burned areas compared with dense shrub cover, although these differences can be limited to the first two months after the fire (i.e., [54]). Our results, indicated that these differences in the long-term have been declining in the plot burned once in 1991 (recover with vegetation) and that major differences exist in the case of the plot burned twice (in 1993 and 2001), suggesting that the high occurrence of fires hampers the recovery of a dense shrub cover, producing higher losses of suspended and solute sediments. In that respect, Ruiz-Flaño et al. [55] indicated that fire reiterations, to remove shrublands and regenerate pasturelands in the Pyrenees, caused the loss of the fertile soil layer due to soil erosion and increased rock cover on the soil's surface.

The results of our study highlight that runoff volumes and soil losses are low in uncultivated plots, due to the high vegetation cover (see Figure 2). It is well known that vegetation reduces surface runoff and soil erosion (due to the protection to splash) and indirectly improves soil quality (increasing fertility, porosity and the number and stability of the aggregates) which favours water infiltration and decreases surface runoff [12,17,43]. Nevertheless, it should be remembered that plant succession after land abandonment is often a very slow process due to low rainfall values, soil depletion during the cultivation period [56] or human disturbances after land abandonment (overgrazing and fires) [55]. In these cases, soil losses are very high due to soil erosion processes (rills, gullies), producing an irreversible period of soil degradation, which is only possible to solve with expensive restorations measures [57].

In this study, carbonate dissolution is relevant for water chemistry in all land uses, as illustrated by the dominance of Ca^{2+} and HCO_3^- outputs (Figure 4). This result is related to the lithology of the study area (Eocene Flysch), dominated by carbonate rocks and is consistent with other studies in mainly limestone catchments (i.e., [58]). In fact, Walling & Webb [59] noted that Ca^{2+} and HCO_3^- represent the prevailing cations and anions in most waters of the world (see also [60]). Notwithstanding, one of the most relevant results is that the proportion of Ca^{2+} and HCO_3^- greatly increases in the runoff of the lower exporting LULC, thus demonstrating that the most important soil nutrients (Mg^{2+}, K^+, Na^+) are strongly controlled by dense shrub cover (dense shrub and abandoned field plots) and grazed grasslands. Furthermore, the relative contributions of dissolved inorganic nitrogen forms differ among LULC, with N-NO$_3$ being the predominant inorganic nitrogen forms in all land uses, especially in agricultural plots because of the addition of chemical fertilizer, the factor that also explains the losses of organic phosphorous. In any case, the higher presence of phosphorous within the solute outputs in the LULC particularly affected by soil erosion is probably due to the prevailing sheet wash erosion recorded in the experimental plots, since phosphates are strongly attached to soil particles and are carried away with them by overland flow. In fact, soil analyses performed in the same study area by Ruiz-Flaño [61] demonstrated that soil samples taken from the lower part of the fields have a higher P content than those taken from the upper part. This result is also consistent with those obtained in other studies (i.e., [62]).

Standard deviations illustrate the distinct behaviour of solutes and suspended sediment. The former shows very low standard deviations, much lower than those from suspended sediment. This suggests that solutes are always present in the runoff with similar values, regardless of the volume of overland flow, whereas suspended sediment is affected by extremely high variability, given its close dependency on the intensity and volume of precipitation and runoff. Furthermore, the small range in the values of solute concentration than in suspended sediment concentration (Table 2) also suggests the limitations of experimental plots to study solute outputs, since they do not record the decisive effect of rainfall infiltration within the soil; therefore, only the effects of soil surface dissolution processes are included, so that the differences in solute concentration among LULC are relatively low.

Implications for Land Management

Both suspended and solute concentrations, as well as the total outputs, provide relevant information for land management in degraded areas with steep land. The analysis of suspended sediment yield from the AVES reveals that cultivation of steep slopes contributes heavily to soil erosion, with higher soil losses from the cereal, fallow land and active shifting agriculture plots [17,56,63]. In particular, shifting agriculture yielded very high quantities of suspended sediment, explaining why a stone paving completely covered the soil surface in the areas that were cultivated in the past with slash-and-burn practices and the general landscape disturbance in the most marginal mountain areas [62,64]. Conversely, sediment yield from scarcely disturbed areas showed very low erosion rates. These results roughly coincide with the solute losses presented in this study, with a progressive gradation from the less disturbed plots to the cereal and fallow land plots, confirming that annual ploughing and fertilizing are critical factors for solute losses. Those plots that were disturbed in the past and now are affected by plant colonization (burned 1 and abandoned field plots) also show relatively low solute losses, followed by burned plot 2, indicating the positive effects of a dense herbaceous and shrub cover. The burned plot 2 shows a high suspended sediment concentration in the long term (due to the double occurrence of fires), although the progressive plant recovery following the fires and the absence of soil removal tend to record relatively low solute concentrations.

For land management purposes, the results obtained recommend (i) avoiding cultivation on steep slopes and encouraging the abandonment of farmland to enhance recolonization of the soil with herbs and shrubs; this is particularly important in the case of shifting agriculture; (ii) promoting the presence of dense shrub or tree cover in order to control runoff and solute and suspended sediment concentrations; (iii) promoting clearing of gentle and concave hillslopes in order to generate high quality water resources and create a complex mosaic with alternating grazing meadows and shrub areas, enabling extensive stockbreeding and landscape heterogeneity; and (iv) avoiding recurrent fires that contribute to temporarily increase solute and suspended sediment losses and deteriorate the capacity of soils for rapid plant recolonization [65]. These results, together with others on various spatial scales, should help to program land abandonment in areas where cropland withdrawal is expected in the coming decades [66].

5. Conclusions

Runoff water, suspended sediment and solute concentration and loads were monitored in the AVES in nine 30 m^2 experimental plots that represent distinct LULC over a 20-year study period (1991–2011). The results confirm the hypothesis that solute concentrations and loads from traditional and present-day agricultural activities is greater than from dense vegetated areas: dense shrub cover and grazing meadows clearly reduce runoff, sediment yield and solute concentrations and loads. The contrary occurs with agricultural activities (cereal fields, fallow land and shifting agriculture). In addition, the results indicate that the plot burned only once presents high solute losses in the short-term, due to the rapid recovery of vegetation, whereas the plot burned twice showed higher solute loads during the whole study period.

Significant differences among LULC were observed and two groups have been identified related to solute concentration and total sediment yield: (i) in the first group, dense shrub cover, grazing meadow, the plot burned once several years ago (and at present densely colonized with shrubs) and abandoned field record the lowest solute concentrations and export small quantities of dissolved solids and sediment; and (ii) in an intermediate position, the plot burned twice several years ago registered moderate values due to the positive influence of relatively rapid plant colonization and cereal, fallow land and both shifting agriculture (active and abandoned) plots record the highest solute concentrations and yield large quantities of solutes, because of ploughing and poor plant cover. Thus, for solute concentration the ranking of LULC ranges from the less disturbed plots to those where cropping is still fully functioning. These results confirm that farmland abandonment represents a decrease in solute outputs (as in suspended sediment) in areas of steep land and that abandoned fields tend to record similar values to those with dense shrub cover.

Author Contributions: Conceptualization, T.L. and J.M.G.-R.; methodology, T.L., J.M.G.-R., E.N.-R.; formal analysis, E.N.-R.; investigation, E.N.-R., M.K., T.L., J.M.G.-R.; resources, T.L., J.M.G.-R., E.N.-R.; data curation, E.N.-R.; writing—original draft preparation, E.N.-R., T.L., J.M.G.-R.; writing—review and editing, E.N.-R., M.K., T.L., J.M.G.-R.; visualization, E.N.-R., M.K.; supervision, T.L. and J.M.G.-R.; project administration, T.L. and J.M.G.-R.; funding acquisition, T.L. and J.M.G.-R.

Funding: This research was supported by the ESPAS project (CGL2015-65569-R, funded by the MINECO-FEDER). The "Geomorphology and Global Change" (E02_17E) research group was financed by the Aragón Government and the European Social Fund (ESF-FSE). Makki Khorchani is working with an FPI contract (BES-2016-077992) from the Spanish Ministry of Economy and Competitiveness associated to the ESPAS project.

Conflicts of Interest: The authors declare no conflict of interest.

References

1. Geissen, V.; Mol, H.; Klumpp, E.; Umlauf, G.; Nadal, M.; van der Ploeg, M.; van de Zee, S.E.; Ritsema, C.J. Emerging pollutants in the environment: A challenge for water resource management. *Int. Soil Water Conserv. Res.* **2015**, *3*, 57–65. [CrossRef]
2. Viviroli, D.; Dürr, H.H.; Messerli, B.; Meybeck, M.; Weingartner, R. Mountains of the world-water towers for humanity: Typology, mapping and global significance. *Water Resour. Res.* **2007**, *43*, W07447. [CrossRef]
3. De Jong, C.; Lawler, D.; Essery, R. Mountain hydroclimatology and snow seasonality and hydrological change in mountain environments. *Hydrol. Process* **2009**, *23*, 955–961. [CrossRef]
4. Kilic, S.; Evrendilek, F.; Berberoglu, S.; Demirkesen, A.C. Environmental monitoring of land-use and land-cover changes in a Mediterranean Region of Turkey. *Environ. Monit. Assess.* **2006**, *114*, 157–168. [CrossRef]
5. Rico-Amorós, A.M.; Olcina-Cantos, J.; Saurí, D. Tourist land use patterns and water demand: Evidence from the Western Mediterranean. *Land Use Pol.* **2009**, *26*, 493–501. [CrossRef]
6. Jlassi, W.; Nadal-Romero, E.; García-Ruiz, J.M. Modernization of new irrigated lands in a scenario of increasing water scarcity: From large reservoirs to small ponds. *Cuad. Investig. Geogr.* **2016**, *42*, 233–259. [CrossRef]
7. García-Ruiz, J.M.; López-Moreno, J.I.; Vicente-Serrano, S.M.; Lasanta, T.; Beguería, S. Mediterranean water resources in a global change. *Earth Sci. Rev.* **2011**, *105*, 121–139. [CrossRef]
8. Giorgi, F. Climate change hot-spots. *Geophys. Res. Lett.* **2006**, *33*, L08707. [CrossRef]
9. López-Moreno, J.I. Recent variations of snowpack depth in the Central Spanish Pyrenees. *Arct. Antarct. Alp. Res.* **2005**, *37*, 253–260. [CrossRef]
10. Nogués-Bravo, D.; Lasanta, T.; López-Moreno, J.I.; Araújo, M.B. Climate warming in Mediterranean mountains during the 21st century. *AMBIO* **2008**, *37*, 280–285. [CrossRef]
11. Vicente-Serrano, S.M.; Lasanta, T.; Romo, A. Analysis of spatial and temporal evolution of vegetation cover in the Spanish Central Pyrenees: Role of human management. *Environ. Manag.* **2005**, *34*, 802–818. [CrossRef] [PubMed]
12. García-Ruiz, J.M.; Lana-Renault, N. Hydrological and erosive consequences of farmland abandonment in Europe, with special reference to the Mediterranean region—A review. *Agric. Ecosyst. Environ.* **2011**, *140*, 317–338. [CrossRef]
13. Beguería, S.; López-Moreno, J.I.; Lorente, A.; Seeger, M.; García-Ruiz, J.M. Assessing the effect of climate oscillations and land-use changes on streamflow in the Central Spanish Pyrenees. *AMBIO* **2003**, *32*, 283–286. [CrossRef]
14. López-Moreno, J.I.; Latron, J. Influence of forest canopy on snow distribution in a temperate mountain range. *Hydrol. Process* **2008**, *22*, 117–1266. [CrossRef]
15. Cosandey, C.; Andréassian, V.; Martin, C.; Didon-Lescot, J.F.; Lavabre, J.; Folton, N.; Mathys, N.; Richard, D. The hydrological impact of the Mediterranean forest: A review of French research. *J. Hydrol.* **2005**, *301*, 235–249. [CrossRef]

16. López-Moreno, J.I.; Zabalza, J.; Vicente-Serrano, S.M.; Revuelto, J.; Gilaberte, M.; Azorin-Molina, C.; Morán-Tejeda, E.; García-Ruiz, J.M.; Tague, C. Impact of climate and land use change on water availability and reservoir management: Scenarios in the Upper Aragon River, Spain Pyrenees. *Sci. Total Environ.* **2014**, *493*, 1222–1231. [CrossRef]

17. Nadal-Romero, E.; Lasanta, T.; García-Ruiz, J.M. Runoff and sediment yield from land under various uses in a Mediterranean mountain area: Long-term results from an experimental station. *Earth Surf. Process. Landf.* **2013**, *38*, 346–355. [CrossRef]

18. San-Miguel-Ayanz, J.; Moreno, J.M.; Camia, A. Analysis of large fires in European Mediterranean landscapes: Lessons learned and perspectives. *For. Ecol. Manag.* **2013**, *294*, 11–22. [CrossRef]

19. Sitzia, T.; Semenzato, P.; Trentanovi, G. Natural reforestation is changing spatial patterns of rural mountain and hill landscapes: A global overview. *For. Ecol. Manag.* **2010**, *259*, 1354–1362. [CrossRef]

20. San Román Sanz, A.; Fernández, C.; Mouillot, F.; Ferrat, L.; Istria, D.; Pasqualini, V. Long-term forest dynamics and land-use abandonment in the Mediterranean mountains, Corsica, France. *Ecol. Soc.* **2013**, *18*, 38. [CrossRef]

21. Molinillo, M.; Lasanta, T.; García-Ruiz, J.M. Managing degraded landscape after farmland abandonment in the Central Spanish Pyrenees. *Environ. Manag.* **1997**, *21*, 587–598. [CrossRef]

22. San Emeterio, L.; Múgica, L.; Ugarte, M.D.; Goicoa, T.; Canals, R.M. Sustainability of traditional pastoral fires in highlands under global change: Effects on soil function and nutrient cycling. *Agric. Ecosyst. Environ.* **2016**, *225*, 155–163. [CrossRef]

23. Morán-Ordoñez, A.; Buster, R.; Suárez-Seoane, S.; de Luis, E.; Calvo, L. Temporal changes in socio-ecological systems and their impact on ecosystems services at different governance scales: A case study of heathlands. *Ecosystems* **2013**, *16*, 765–782. [CrossRef]

24. Bernués, A.; Rodríguez-Ortega, T.; Ripoll-Bosch, R.; Alfnes, F. Socio-cultural and economic valuation of ecosystem services provided by Mediterranean agroecosystems. *PLoS ONE* **2014**, *9*, e102479. [CrossRef]

25. Fernandes, P.M.; Davies, G.M.; Ascoli, D.; Fernández, C.; Moreira, F.; Rigolot, E.; Stoff, C.R.; Vega, J.A.; Molina, D. Prescribed burning in southern Europe: Developing fire management in a dynamic landscape. *Front. Ecol. Environ.* **2013**, *11*, e4–e14. [CrossRef]

26. Lasanta, T.; Nadal-Romero, E.; Errea, M.P.; Arnáez, J. The effects of landscape conservation measures in changing landscape patterns: A case study in Mediterranean mountain. *Land Degrad. Dev.* **2016**, *27*, 373–386. [CrossRef]

27. Lasanta, T.; Nadal-Romero, E.; García-Ruiz, J.M. Clearing shrubland as a strategy to encourage extensive livestock farming in the Mediterranean mountains. *Cuad. Investig. Geogr.* **2019**, *45*. [CrossRef]

28. Alados, C.L.; Saiz, H.; Nuche, P.; Gartzia, M.; Komac, B.; De Frutos, Á.; Pueyo, Y. Clearing vs. burning for restoring Pyrenean grasslands after shrub encroachment. *Cuad. Investig. Geogr.* **2019**, *45*. [CrossRef]

29. Le Houérou, H.N. Agroforestry and sylvopastoralism to combat land degradation in the Mediterranean basin: Old approaches to new problems. *Agric. Ecosyst. Environ.* **1990**, *33*, 99–109. [CrossRef]

30. Shalaby, A.; Tateishi, R. Remote sensing and GIS for mapping and monitoring land cover and land-use changes in the Northwestern coastal zone of Egypt. *Appl. Geogr.* **2007**, *27*, 28–41. [CrossRef]

31. Gispert, M.; Pardini, G.; Emran, M.; Doni, S.; Masciandaro, G. Seasonal evolution of soil organic matter, glomalin and enzymes and potential for C storage after land abandonment and renaturalization processes in soils of NE Spain. *Catena* **2018**, *162*, 402–413. [CrossRef]

32. Lizaga, I.; Quijano, L.; Gaspar, L.; Ramos, M.C.; Navas, A. Linking land use changes to variation in soil properties in a Mediterranean mountain agroecosystem. *Catena* **2019**, *172*, 516–527. [CrossRef]

33. Robledano-Aymerich, F.; Romero-Díaz, A.; Belmonte-Serrato, F.; Zapata-Pérez, V.M.; Martínez-Hernández, C.; Martínez-López, V. Ecogeomorphological consequences of land abandonment in semiarid Mediterranean areas: Integrated assessment of physical evolution and biodiversity. *Agric. Ecosyst. Environ.* **2014**, *197*, 222–242. [CrossRef]

34. Cerdà, A.; Rodrigo-Comino, J.; Novara, A.; Brevik, E.C.; Vaezi, A.R.; Pulido, M.; Giménez-Morera, A.; Keesstra, S.D. Long-term impact of rainfed agricultural land abandonment on soil erosion in the Western Mediterranean basin. *Prog. Phys. Geogr.* **2018**, *42*, 202–219. [CrossRef]

35. Trudgill, S.T. Introduction. In *Solute Processes*; J Wiley: Chichester, UK, 1986; pp. 1–14.

36. Smaling, E.M.A.; Oenema, O. Estimating nutrient balances in agro-ecosystems at different spatial scales. In *Methods for Assessment of Soil Degradation*; CRC Press: Boca Raton, FL, USA, 1997; pp. 229–252.

37. Han, G.; Li, F.; Tan, Q. Effects of land use on water chemistry in a river draining karst terrain, southwest China. *Hydrol. Sci. J.* **2014**, *59*, 1063–1073. [CrossRef]

38. Germer, S.; Neill, C.; Vetter, T.; Chaves, J.; Krusche, A.V.; Elsenbeer, H. Implications of long-term land-use change for the hydrology and solute budget of small catchments in Amazonia. *J. Hydrol.* **2009**, *364*, 349–363. [CrossRef]

39. Figuereido, R.O.; Markewitz, D.; Davidson, E.A.; Schuler, A.E.; Watrin, O.; de Souza Silva, P. Land-use effects on the chemical attributes of low-order streams in the Eastern Amazon. *J. Geophys. Res.* **2010**, *115*, G04004. [CrossRef]

40. Morales-Martín, L.; Wheater, H.; Lindenschmidt, K.E. Potential changes of annual averaged nutrient export in the South Saskatchewan River Basin under climate and land use change scenarios. *Water* **2018**, *10*, 1438. [CrossRef]

41. Llorens, P.; Queralt, I.; Plana, F.; Gallart, F. Studying solute and particulate sediment transfer in a small Mediterranean mountainous catchment subject to land abandonment. *Earth Surf. Process. Landf.* **1997**, *22*, 1027–1035. [CrossRef]

42. Outeiro, L.; Úbeda, X.; Farguell, J. The impact of agricultura on solute and suspended sediment load on a Mediterranean watershed after intense rainstorms. *Earth Surf. Process. Landf.* **2010**, *35*, 549–560.

43. Nadal-Romero, E.; Lana-Renault, N.; Serrano-Muela, P.; Regüés, D.; Alvera, B.; García-Ruiz, J.M. Sediment balance in four catchments with different land cover in the Central Spanish Pyrenees. *Z. Geomorphol.* **2012**, *56*, 147–168. [CrossRef]

44. Oliveira, J.S.; da Cunha Bustamante, M.M.; Markewitz, D.; Krusche, A.V.; Ferreira, L.G. Effects of land cover on chemical characteristics of streams in the Cerrado region of Brazil. *Biogeochemistry* **2011**, *105*, 75–88.

45. Ferreira, C.S.S.; Walsh, R.P.D.; Costa, M.; Oliveira Alves, C.; Ferreira, A.J.D. Dynamics of Surface water quality driven by distinct urbanization patterns and storms in a Portuguese peri-urban catchment. *J. Soils Sediments* **2016**, *16*, 2606–2621. [CrossRef]

46. Merchán, D.; Luquin, E.; Hernández-García, I.; Campo-Bescós, M.A.; Giménez, R.; Casalí, J.; Del Valle de Lersundi, J. Dissolved solids and suspended sediment dynamics from five small agricultural watersheds in Navarre, Spain: A 10-year study. *Catena* **2019**, *173*, 114–130. [CrossRef]

47. Messerli, B.; Vivirioli, D.; Weingartner, R. Mountains of the World-vulnerable water towers for the 21st century. *AMBIO Spec. Rep.* **2004**, *13*, 29–34.

48. Boix-Fayos, C.; Martínez-Mena, M.; Arnau-Rosalén, E.; Calvo-Cases, A.; Castillo, V.; Albaladejo, J. Measuring soil erosion by field plots: Understanding the sources of variation. *Earth Sci. Rev.* **2006**, *78*, 267–285. [CrossRef]

49. Risking, S.H.; Neill, C.; Jankowski, K.; Krusche, A.V.; McHorney, R.; Elsenbeer, H.; Macedo, M.N.; Nunes, D.; Porder, S. Solute and sediment export from Amazon forest and soybean headwater streams. *Ecol. Appl.* **2017**, *27*, 193–207. [CrossRef]

50. Food and Agriculture Organization (FAO). *Water Pollution from Agriculture: A Global Review*; Food and Agriculture Organization of the United Nations: Rome, Italy; The International Water Management Institute on Behalf of the Water Land and Ecosystems Research Program: Colombo, Sri Lanka, 2017.

51. Lasanta, T.; Cerdà, A. Long-term erosional responses after fire in the Central Spanish Pyrenees. 2. Solute release. *Catena* **2005**, *60*, 81–100. [CrossRef]

52. Zavala, L.M.; De Celis, R.; Jordán, A. How wildfires affect soil properties. *A brief review. Cuad. Investig. Geogr.* **2014**, *40*, 311–331.

53. Machado, A.; Serpa, D.; Ferreira, R.V.; Rodríguez-Blanco, M.L.; Pinto, R.; Nunes, M.; Cerqueira, M.A.; Keizer, J.J. Cation export by overland flow in a recently burnt forest area in north-central Portugal. *Sci. Total Environ.* **2015**, *524–525*, 201–212. [CrossRef]

54. Ferreira, R.V.; Serpa, D.; Machado, A.I.; Rodríguez-Blanco, M.L.; Santos, L.F.; Taboada-Castro, M.T.; Cerqueira, M.A.; Keizer, J.J. Short-term nitrogen losses by overland flow in a recently burnt forest area in north-central Portugal: A study at micro-plot scale. *Sci. Total Environ.* **2016**, *572*, 1281–1288. [CrossRef]

55. Ruiz-Flaño, P.; García-Ruiz, J.M.; Ortigosa, L. Geomophological evolution of abandoned fields. A case study in the Central Pyrenees. *Catena* **1992**, *19*, 301–308. [CrossRef]

56. Lasanta, T.; Nadal-Romero, E.; Errea, M.P. The footprint of marginal agriculture in the Mediterranean mountain landscape: An analysis of the Central Spanish Pyrenees. *Sci. Total Environ.* **2017**, *599–600*, 1823–1836. [CrossRef]

57. Pueyo, Y.; Alados, C.L.; García-Ávila, B.; Kefi, S.; Maestro, M.; Rietberk, M. Comparing direct abiotic amelioration and facilitation as tools for restoration on semi-arid grassland. *Restor. Ecol.* **2009**, *17*, 908–916. [CrossRef]
58. Llorens, P.; Gallart, F. Small basin response in a Mediterranean mountainous abandoned farming area: Research design and preliminary results. *Catena* **1992**, *19*, 309–320. [CrossRef]
59. Walling, D.E.; Webb, B.W. Solutes in river systems. In *Solute Processes*; J Wiley: Chichester, UK, 1986; pp. 251–327.
60. Crabtree, R.W.; Trudgill, S.T. Hydrochemical Budgets for a Magnesium Limestone catchment in low land England. *J. Hydrol.* **1984**, *74*, 67–79. [CrossRef]
61. Ruiz-Flaño, P. *Procesos de Erosión en Campos Abandonados del Pirineo*; Geoforma Ediciones: Logroño, Spain, 1993; 220p.
62. Casali, J.; Giménez, R.; Díez, J.; Álvarez-Mozos, J.; Del Valle de Lersundi, J.; Goñi, M.; Campo, M.A.; Chahor, Y.; Gastesi, R.; López, J. Sediment production and water quality of watersheds with contrasting land use in Navarre (Spain). *Agric. Water Manag.* **2010**, *97*, 1683–1694. [CrossRef]
63. Lasanta, T.; Beguería, S.; García-Ruiz, J.M. Geomorphic and hydrological effects of traditional shifting agriculture in a Mediterranean mountain area, Central Spanish Pyrenees. *Mt. Res. Dev.* **2006**, *26*, 146–152. [CrossRef]
64. García Ruiz, J.M.; Lasanta, T. El Pirineo aragonés como paisaje cultural. *Pirineos* **2018**, e038. [CrossRef]
65. Pausas, J.G.; Llovet, J.; Rodrigo, A.; Vallejo, R. Are wildfire a disaster in the Mediterranean basin? A review. *Int. J. Wildland Fire* **2008**, *17*, 6. [CrossRef]
66. Verburg, P.H.; Overmars, K.P. Combining top-down and bottom-up dynamics in land use modeling: Exploring the future of abandoned farmlands in Europe with the Dyna-CLUE model. *Landsc. Ecol.* **2009**, *24*, 1167–1181. [CrossRef]

Article

Investigation on Farmland Abandonment of Terraced Slopes Using Multitemporal Data Sources Comparison and Its Implication on Hydro-Geomorphological Processes

Giacomo Pepe [1,2,*], Andrea Mandarino [1,2], Emanuele Raso [2,3], Patrizio Scarpellini [4], Pierluigi Brandolini [1] and Andrea Cevasco [1]

[1] Department of Earth, Environmental and Life Sciences, University of Genova, Corso Europa 26, 16132 Genova, Italy
[2] Geoscape Soc. Coop., Geo-Environmental Consulting, Via Varese 2, 16122 Genova, Italy
[3] Department of Earth Sciences, Environment and Resources, Federico II University of Napoli, Complesso Universitario Monte Sant'Angelo, Building L1, Via Cinthia, 80126 Napoli, Italy
[4] Ente Parco Nazionale delle Cinque Terre, Via Discovolo snc, Manarola, 19017 Riomaggiore, Italy
* Correspondence: giacomo.pepe@unige.it; Tel.: +39-01-0353-8206

Received: 9 July 2019; Accepted: 25 July 2019; Published: 2 August 2019

Abstract: This paper presents a quantitative multi-temporal analysis performed in a GIS environment and based on different spatial information sources. The research is aimed at investigating the land use transformations that occurred in a small coastal terraced basin of Eastern Liguria from the early 1950s to 2011. The degree of abandonment of cultivated terraced slopes together with its influence on the distribution, abundance, and magnitude of rainfall-induced shallow landslides were accurately analysed. The analysis showed that a large portion of terraced area (77.4%) has been abandoned over approximately sixty years. This land use transformation has played a crucial role in influencing the hydro-geomorphological processes triggered by a very intense rainstorm that occurred in 2011. The outcomes of the analysis revealed that terraces abandoned for a short time showed the highest landslide susceptibility and that slope failures affecting cultivated zones were characterized by a lower magnitude than those which occurred on abandoned terraced slopes. Furthermore, this study highlights the usefulness of cadastral data in understanding the impact of rainfall-induced landslides due to both a high spatial and thematic accuracy. The obtained results represent a solid basis for the investigation of erosion and the shallow landslide susceptibility of terraced slopes by means of a simulation of land use change scenarios.

Keywords: agricultural terraces; cadastral map; cadastral register; farmland abandonment; GIS; land use changes; shallow landslides; terraced slopes

---

1. Introduction

Temporal variation in land use and land cover (LULC) represents one of the main environmental factors controlling the occurrence of natural phenomena like landslides [1,2] and floods [3–6]. In particular, LULC can influence the distribution, abundance, and magnitude of the hydrological processes of shallow landsliding. In the last twenty years, numerous studies have considered the effects of changing land use scenarios in the analysis of landslide susceptibility [7–12], hazard [13,14], and risk [15,16].

LULC modifications directly or indirectly induced by human activity, such as extensive deforestation practices or road and constructions built in hazardous zones, often produce the most severe landslide-related consequences. For instance, forest logging can decrease the positive hydro-mechanical effects of vegetation (e.g., soil reinforcements by roots, reduction of rainfall infiltration and pore water pressure by evapotranspiration processes) [17,18], causing an increase in the proneness of slopes to erosion phenomena like mass movements and runoff [19,20]. Among the indirect human-related land use modifications, the abandonment of farming areas can be considered by far one of the most relevant [21–27]. In Europe, due to the significant social and economic changes which occurred after the Second World War [28], many hilly and mountainous regions have experienced severe and accelerated slope degradation issues led by farmland abandonment [29–34]. Slope degradation in conjunction with extreme rainfall can produce a significant growth in susceptibility to erosion and landslide phenomena [35–42]. Moreover, abandoned slopes may become source of risk scenarios when located in the proximity of urban areas [43,44].

The negative effect of farmland abandonment on slope stability can be especially important in terraced landscapes. It is widely known from technical literature that terraces concur to improve slope stability by reducing the overall slope gradient and by regulating water infiltration and runoff [45–50]. However, due to agricultural abandonment, the basic components of terraced systems, namely dry-stone walls and complementary drainage structures, are no longer adequately maintained and managed, causing the efficiency of the hydrological and retaining functions to decrease. Accordingly, agricultural terraces become more vulnerable and therefore highly susceptible to collapses and failures [51–55]. Many studies addressed that the degree of abandonment is a crucial factor in regulating the susceptibility of terraced slopes to be affected by rainfall-induced landslides [11,56–59].

Unlike other environmental landslide predisposing factors like geology, geomorphology or soil types, LULC is not a static factor since it may undergo significant variations over time [1]. Consequently, for landslide hazard and risk assessment purposes this causal factor should be regularly updated according to LULC changes in a specific area. The analysis of past land use settings and modifications is therefore essential to assess landslide susceptibility of an area as well as to estimate the future occurrence of slope failures. As reported by some authors [15,60], in order to perform a reliable analysis of land use changes it is essential to correctly determine where these modifications occurred and their rates. Nowadays, the availability of digital tools for managing spatial information such as geographical information systems (GIS) together with remote sensing techniques represents a very useful way of investigating LULC changes over large areas. These methods allow for the production of multi-temporal LULC maps and LULC change maps that are fundamental data sets in the framework of both landslide zoning and GIS-based modelling of landslide susceptibility and hazard [1,2]. In order to create past LULC thematic maps of an area, the interpretation of historical aerial photos is usually performed. However, also ancillary data can be a precious data source. Old cadastral maps, for example, are often the only available source of information about past land use setting. Cadastral data can cover several time frames often providing information at a very detailed scale. There are several examples about the use of cadastral maps to analyse spatial and temporal variation in land use. Nevertheless, these studies are usually focused on analysing socioeconomic or political factors behind land use changes and their consequences on landscape, ecology and biodiversity [61–66]. Moreover, cadastral databases frequently represent basic elements in landslide risk analysis [1], even though very few researches [67] have used cadastral data to understand the impact of landslides in response to land use transformations.

Terraced slopes are one of the most relevant morphological features of the Ligurian landscape (north-western Italy). This region is predominantly hilly and mountainous and since the early centuries of the Middle Ages wide extensions of natural slopes were terraced to allow cultivation practices. Nowadays, a significant percentage of traditional agricultural terraces is currently abandoned due to accelerated socioeconomic transformations occurred during the last century [68,69].

This paper describes the use of different multitemporal spatial information aimed at quantifying land use modifications affecting agricultural terraced slopes in a small coastal catchment located within the Cinque Terre National Park (eastern Liguria, north-western Italy). The selected study area is indicative of the farmland abandonment spread throughout the entire Liguria region. Historical

LULC maps were derived from old cadastral maps and aerial photographs of the study area and they were compared with the current land use conditions to make an accurate estimate of the degree of abandonment of terraces over the last sixty years. Subsequently, the current spatial pattern of the land use conditions of terraced slopes was analysed with respect to the distribution of the shallow landslides triggered by a very intense rainfall event occurred on 25 October 2011 [11,70]. The objective of this study is twofold. On the one hand, the aim is to verify the usefulness of cadastral data in understanding the impact of rainfall-induced landslides on terraced slopes. On the other hand, the aim is to further improve the knowledge about the relationship between land abandonment and shallow landslide processes.

2. General Setting of the Study Area

The study area is located on the easternmost sector of Liguria (northwestern Italy) and corresponds to the Vernazza catchment, the widest of several small coastal basins of the Cinque Terre area (Figure 1a). The study catchment extends for approximately 5.8 km^2 and from a geological point of view it is included within a segment of the Northern Apennine that is characterized by sedimentary rocks belonging to three tectonic units (bottom to top): Tuscan Nappe, Marra unit, and Canetolo complex [71]. The Marra Unit and the Canetolo complex entirely crop out along a NW-SE oriented stretch of land that occupies the middle portion of the basin (Figure 1b). The first one is prevalently composed of marls and siltstones (Pignone Marls Fm.) while the second one respectively includes assemblages of prevailing shales with subordinate limestones and silty sandstones (Canetolo Shales and Limestones Fm.), marly limestones and calcarenitic turbidites (Groppo del Vescovo Limestones Fm.) and fine-grained sandstone turbidites (Ponte Bratica Sandstones Fm.). On the other hand, the Tuscan Nappe crops out in the majority of the basin area and chiefly consists of typical turbidites made up of sandstones and siltstones (Macigno Fm.). Overall, rock formations are arranged in a complex structural setting characterized by a large, SW-verging, overturned antiform fold and by multiple sets of tectonic discontinuities associated to the Apennine orogeny and to the Plio-Quaternary tectonic up lift [72,73].

As a result of the peculiar geo-structural setting, the study area shows the typical morphology of mountainous regions, in spite of the proximity to the sea [74]. The elevations increase rapidly moving from the coastline to the inland territory. The maximum altitudes are widespread all around the main watershed and are represented by the peaks of Mt. S. Croce (622 m), Mt. Gaginara (772 m) and Mt. Malpertuso (815 m) (Figure 1b). The inner territory of the basin is carved by numerous small and narrow V-shaped valleys drained by a dense dendritic channel network with steep profile while the shoreline is mostly characterized by rocky cliffs [73–75]. Generally, the streams are short and show an ephemeral hydrological regime. The final reach of the main stream was diverted in the past and currently it flows partly culverted and partly through a tunnel for a total length of over 150 m. The small V-shaped valleys are limited by steep to very steep slopes (Figure 1c,d) covered by thin veneers (thicknesses ranging between 1 and 2.5 m on average) of eluvial-colluvial deposits [11,76] that in the past were largely reworked by local inhabitants in order to build agricultural terraces, mainly cultivated with vineyards and olive groves [74,77]. The resulting terraced landscape is now a worldwide-known tourist attraction declared since 1997 as "UNESCO World Heritage" for its high scenic and cultural value and since 1999 as a national park due to its environmental and naturalistic relevance.

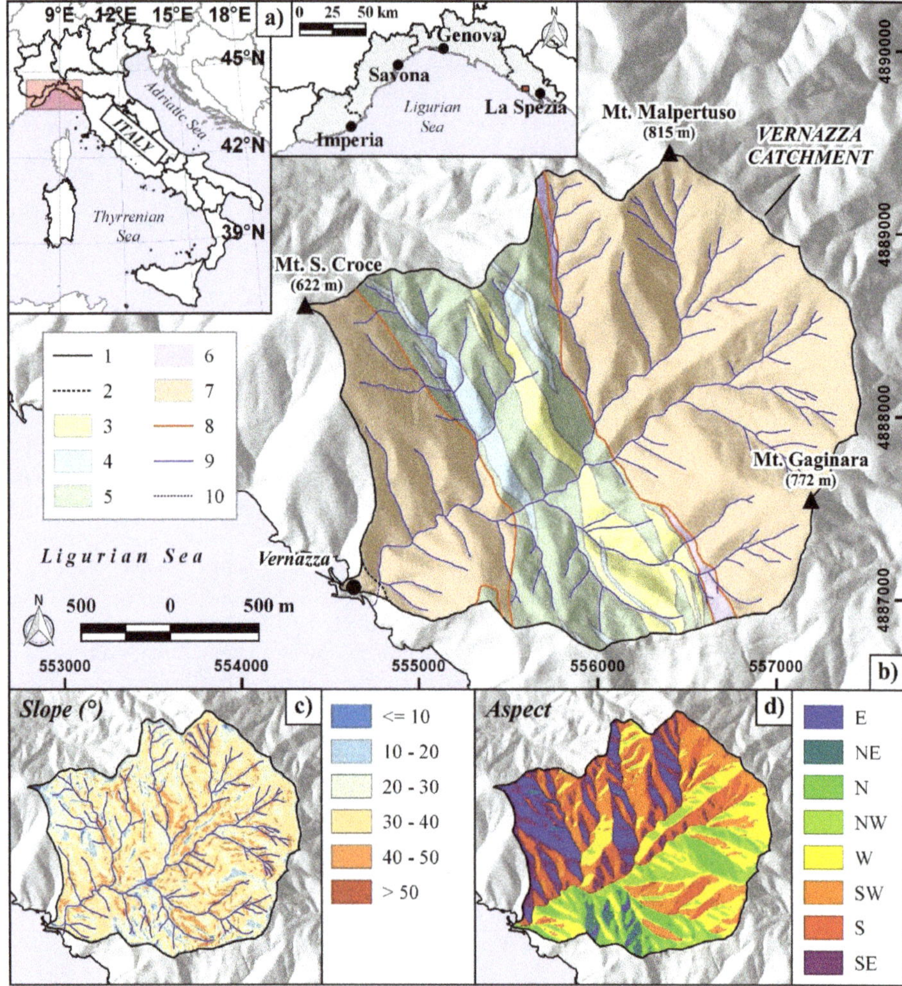

Figure 1. (**a**) Location of the study area; (**b**) Geologic map of the Vernazza basin (redrawn and modified from [71]): 1, Vernazza catchment divide; 2, downstream-most limit of the investigated catchment 3, Ponte Bratica sandstones Fm.; 4, Groppo del Vescovo limestones Fm.; 5, Canetolo shales and limestones Fm.; 6, T. Pignone marls Fm.; 7 Macigno Fm.; 8, main tectonic features (e.g., direct and reverse faults, overthrusts); 9, hydrographic network; 10, tunnelled reach of the main stream; (**c**) slope map of the Vernazza basin; (**d**) aspect map of the Vernazza basin (derived from 5 m cell-size DEM, source: Geoportale Regione Liguria).

The current land use setting was analyzed through recent researches [78]. Natural areas, prevalently including woods and scrub, cover approximately 50% of the study area and are mostly distributed along the upper portions of the catchment. The remaining 50% of the basin, encompassing the middle and lower zones of the catchment, is entirely covered by terraced slopes characterized by different land uses and degrees of abandonment. However, it is relevant to note that only about 8% of terraced slopes is still cultivated at present time [78]. Urban areas represent a very small portion of the basin and they mainly coincide with the Vernazza village, which is settled on the downstream portion of the valley floor.

The morphological features deeply influence the local climate setting. The orientation of the main watershed parallel and very close (approximately 3 km) to the coastline produces a topographic effect that locally increases the effects of the typical Mediterranean climate. The average annual temperature is about 15 °C while the average annual precipitation is around 1000 mm [35]. Generally, summers are warm and dry whereas winters are somewhat mild. Rainfalls are on average more abundant between the end of summer and the beginning of winter, when very intense and strongly localized rainstorms can often affect this coastal sector of the Liguria Region [35,79]. On 25 October 2011, one of these extreme rainfall events seriously hit the western sector of the Cinque Terre area and especially the Vernazza catchment. Widespread erosion processes were triggered together with several hundred shallow landslides often evolving rapidly into flow-like ones [11].

3. Materials and Methods

This research consists of a quantitative multi-temporal analysis performed in a GIS environment. This work has been carried out in three steps (Figure 2). In the first step, historical and recent aerial photos have been used as mapping basis to obtain LULC data of the study area in 1954 and 2011, respectively.

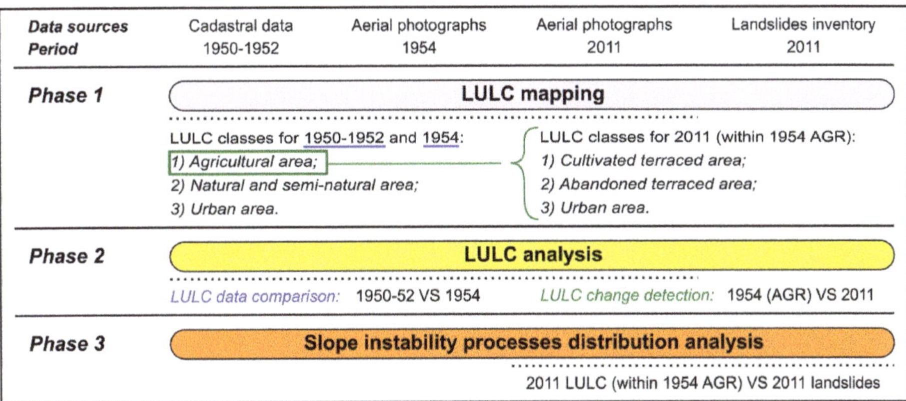

Figure 2. Workflow sketch summarizing the considered data sources and the sequence of performed methodological phases. The dotted lines refer to the data sources and indicate which of them are involved in each phase. Blue- and green-highlighted elements in Phase 1 refer to the Phase 2 components; LULC, land use and land cover; AGR, agricultural area.

In the second step, the 1954 LULC map and the cadastral maps dating back to the early 1950s were brought into comparison to analyze whether cadastral data were geometrically accurate and thematically comparable. Subsequently, an overlay analysis among the 1954 LULC map and the recent land use setting have been performed to investigate the farmland abandonment of terraced slopes. These analyses did not take into account the downstream-most small portion of the catchment corresponding to the historical center of Vernazza since already totally urbanized in the 1950s. Eventually, in the third step the spatial distribution of the shallow landslides triggered by the 25 October 2011 rainstorm was analysed with respect to LULC changes occurred in agricultural areas of 1954.

3.1. Data Sources

This research is based on several types of data sources providing information on both past and recent LULC settings and hydro-geomorphological processes that affected the study area (Figure 2). To map the past LULC, the aerial photographs from the "GAI flight" were integrated in a GIS environment through a georeferencing procedure. This data, approximately at 1:60,000 scale and dating back to 14 September 1954, represents the first stereoscopic photographs depiction covering the whole Italian territory.

To analyze recent LULC, high-resolution orthophotos (ground resolution varying between 3 cm and 50 cm, according to elevation) were used: the original aerial photos were taken by the Air Service of Remote Sensing and Monitoring of Civil Protection of Friuli Venezia Giulia Regional Administration (11 November 2011 flight) just a few days after the catastrophic rainfall event occurred on 25 October 2011. Furthermore, land use information deriving from cadastral maps dating back to early 1950s were considered. These maps were drawn at large scale (from 1:2000 up to 1:500) in order to provide a very detailed spatial information since the associated database reported the classification in terms of LULC for each land parcel. In particular, we used a vector layer made up of polygons representing land parcels which was provided by the Cinque Terre National Park and derived from old cadastral maps.

Lastly, a detailed inventory of rainfall-induced shallow landslides prepared by Cevasco et al. [78] after the 25 October 2011 event was used. It is relevant to point out that in this study only shallow landslides that affected terraced slopes cultivated in 1954 were selected.

3.2. Land Use Mapping

In a GIS environment, the polygons representing land use units were manually mapped at 1:2500 scale through aerial-photos interpretation to obtain the map of the Vernazza catchment depicting the LULC conditions in 1954 (Figure 2). Three main land use classes were visually identified: urban areas (URB), natural and semi-natural areas (NAT) and agricultural areas (AGR). NAT include woodlands, shrubs and zones occupied by herbaceous vegetation. Considering the morphological features of the study area, AGR can be assumed as the extent of the cultivated terraced slopes in 1954.

Subsequently, we produced a land use map using cadastral data temporally referring to the early 1950s. The land use categories used for this second map were the same as the ones adopted in the interpretation of historical aerial photos. Therefore, land use information related to every parcel was reclassified according to the aforementioned three major land use classes, in order to compare heterogeneous data. In detail, the following cadastral parcels were considered as agricultural areas: orchards, arable lands, vineyards, olive groves, pastures and meadows. All patches reported as tall forests and coppice woods were instead included into natural and semi-natural areas while buildings and rural buildings were grouped as urban areas. The parcels belonging to the same class were then dissolved and the areas not covered by parcels (ND), substantially corresponding to state-owned plots of land, were extracted.

Eventually, by using the 2011 high-resolution aerial orthophotos, 1954 agricultural areas (1954 AGR) were classified into three major LULC categories reflecting their recent land use conditions: urban areas, abandoned terraced areas and cultivated terraced areas (Figure 2). In this phase, a number of sub-categories were also defined in order to perform a more detailed analysis. In particular, following the classification adopted by Cevasco et al. [78], abandoned terraced slopes were subdivided into abandoned terraced slopes with poor (ATP) and dense cover (ATD) according to the degree of vegetation development [78]. The former show herbaceous cover or shrubs and can be assumed as abandoned for a short time (less than 25–30 years). The latter have been abandoned for longer time (more than 25–30 years) resulting prevalently occupied by forest tree species. Furthermore, the cultivated zones were divided into two land use sub-classes labelled as areas mostly presenting vineyards (CTV) and olive groves (CTO), respectively.

3.3. Data Processing

The 1954 and the cadastral-based LULC maps were first compared through an overlapping procedure in order to quantify the differences and subsequently to validate the land use information attached to cadastral data. Afterward, to quantify and locate precisely the land use transformations affecting the agricultural areas dated 1954, land use data referred to the 2011 were analyzed by computing the extent of each land use class and sub-class (Figure 3).

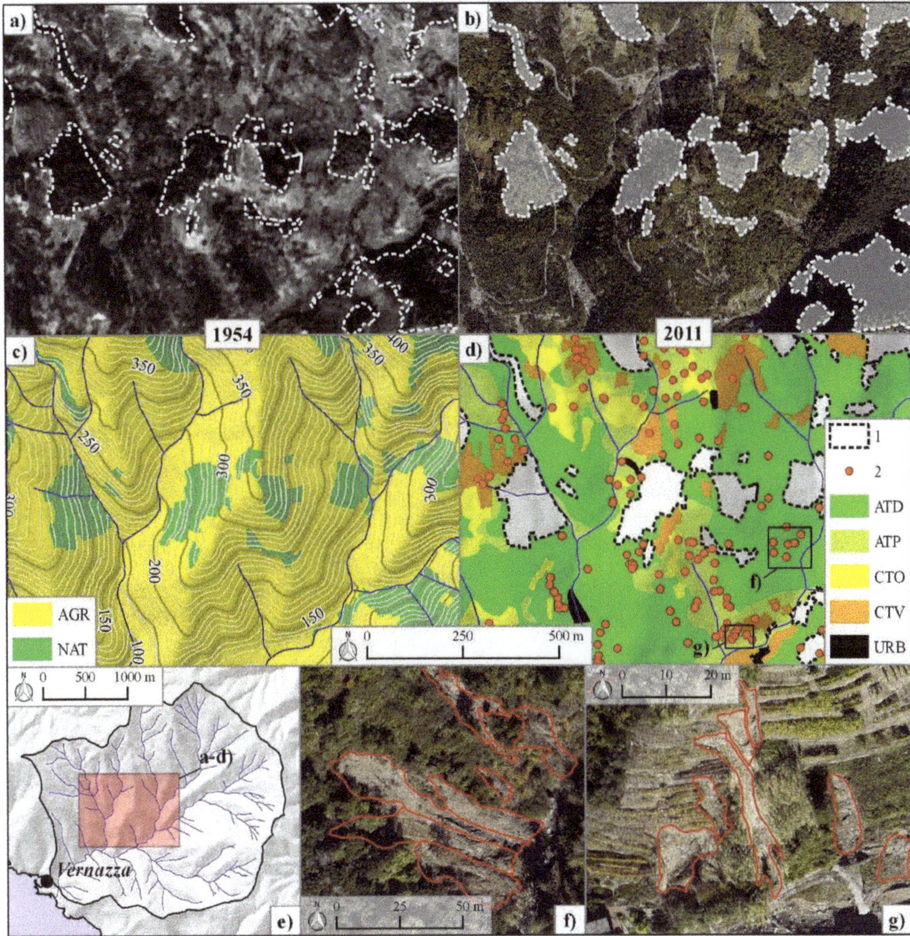

Figure 3. Data sources and derived land use and land cover (LULC) maps: (**a**) Aerial photograph dated 1954; (**b**) Orthophoto dated 2011; (**c**) 1954 LULC map; (**d**) 2011 LULC map in 1954 agricultural area; (**e**) Location of (**a**) to (**d**) sketch maps; (**f**,**g**) examples of landslides triggered on 25 October 2011. The white dashed line in (**a**) separates agricultural and natural and semi-natural area. The white transparent polygons in (**b**) represent natural and semi-natural areas in 1954. The red lines in (**f**,**g**) highlight the perimeter of landslides triggered on abandoned and cultivated terraced slopes, respectively. AGR, agricultural area; NAT, natural and semi-natural area; 1, natural and semi-natural areas in 1954; 2, 2011 landslide sources; ATD, abandoned terraced slope with dense cover; ATP, abandoned terraced slope with poor cover; CTO, cultivated olive grove; CTV, cultivated vineyard; URB, urban area.

Through the last step of the analysis, we investigated the relationships between the outlined land use changes that affected agricultural lands over the last 60 years and the distribution of shallow landslides triggered by the 25 October 2011 rainstorm. Landslide polygons were superimposed on the 2011 land use map and for each LULC category the area affected by landslides was calculated. Furthermore, we digitized a point at the center of every landslide crown to compute the number of slope instability processes that were triggered in each land use category.

Lastly, the landslide area related to each land use class was divided to the total area of that land use type to assess the landslide index (LI).

4. Results

4.1. Land Use Changes Analysis

The LULC map produced through the interpretation of aerial photos shows that in 1954 59.1% of the total basin area (5.8 km^2) was made up of natural and semi-natural areas, followed by cultivated agricultural lands (40.8%) and urban areas (0.1%) (Table 1; Figure 4). The majority of NAT lands almost totally covered the upper portion of the catchment, approximately starting from 400 m of altitude depending on slope aspect, while the remaining ones prevalently extended over the middle portions placed close to the main watershed (Figure 4a). On the other hand, farmland areas (i.e., AGR) diffusely covered the middle part of the catchment whereas they predominated in the lower part. Eventually, the urban areas mainly consisted of the man-made structures belonging to the Vernazza hamlet.

Table 1. Summary of historical land use/land cover (LULC) setting of the Vernazza catchment (AGR, agricultural area; NAT, natural and semi-natural area; URB, urban area; ND, no data).

LULC	1954—From Aerial Photos		Early 1950s—From Cadastral Data	
	%	km^2	%	km^2
NAT	59.1	3.40	59.7	3.44
AGR	40.8	2.35	36.4	2.10
URB	0.1	0.01	0.90	0.05
ND	-	-	3.0	0.17

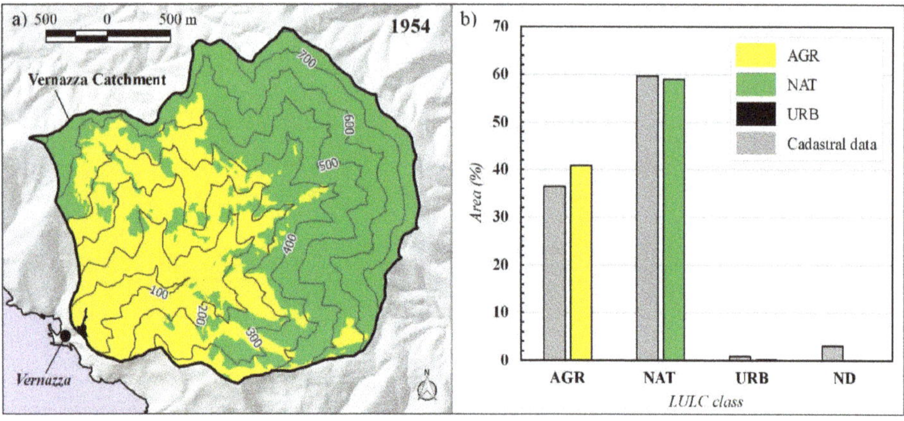

Figure 4. 1954 land use/land cover (LULC) setting of the Vernazza catchment. (**a**) LULC map derived from the 1954 aerial photographs; (**b**) Area in percentage for each LULC class. Colored columns are referred to (**a**) while grey columns indicate the cadastral-derived area for each LULC class; AGR, agricultural area; NAT; natural and semi-natural area; URB, urban area; ND, no data.

The results of the cadastral-based land use mapping lead to a very similar LULC setting (Table 1). In fact, the study catchment showed a prevalence of NAT (59.7%), followed by AGR (36.4%) and lastly by urban areas (0.9%) while a remaining 3% of the basin resulted substantially occupied by state-owned areas (ND) (Table 1; Figure 4b). By comparing the 1954 LULC map with the one obtained through cadastral data, a very slight difference was detected as highlighted by an overlapping degree of approximately 96%. Moreover, it is interesting to note that cadastral maps showed a high land-property fragmentation within the study basin, where approximately 7200 parcels were identified with an average parcel size of about 775 m^2. However, the density of cadastral parcels resulted further higher in agricultural areas, where the average parcel size was 430 m^2. Therefore, cadastral data allow to obtain spatial information at higher detail, greatly improving land use information obtained from aerial-photo interpretation. In fact, according to cadastral database, natural and seminatural areas can be mainly subdivided into tall forests (71%) and coppice woods (29%). On the other hand, agricultural lands consist of pastures and meadows (13%), orchards/arable lands (23.5%), olive groves (24%) and vineyards (39.5%). These results highlight the importance of agriculture activities in the early 1950s and they make possible analysis of both the distribution and extent of cultivated terraced slopes in the study area.

The main changes experienced between 1954 and 2011 by agricultural areas are summarized in Table 2 and Figure 5. It is relevant to note that over approximately sixty years a large portion of agricultural terraced slopes (77.4%) has been abandoned while only a small percentage (21.1%) remained cultivated. In detail, 19% of cultivated terraced slopes have become ATP while 58.4% turned into ATD (Table 2; Figure 5b). Terraced slopes that have been kept cultivated mainly consist of CTV (15.3%) and CTO (5.8%) whereas URB category (1.5%) includes anthropic structures due to urban sprawl of the Vernazza hamlet together with the main rural building settlements distributed in the inland portions of the catchment. It is interesting to point out that cultivated terraces correspond to about 9% of the whole study basin. Currently, such agricultural lands are scattered in small patches, often located in the proximity of roads and/or rural settlements.

Table 2. Summary of land use/land cover (LULC) setting of the 1954 agricultural area in 2011 (ATD, abandoned terraced slope with dense cover; ATP, abandoned terraced slope with poor cover; CTO, cultivated olive grove; CTV, cultivated vineyard; URB, urban area).

LULC	2011—From Aerial Photos	
	%	km²
ATD	58.4	1.38
ATP	19.0	0.45
CTO	5.8	0.14
CTV	15.3	0.36
URB	1.5	0.03

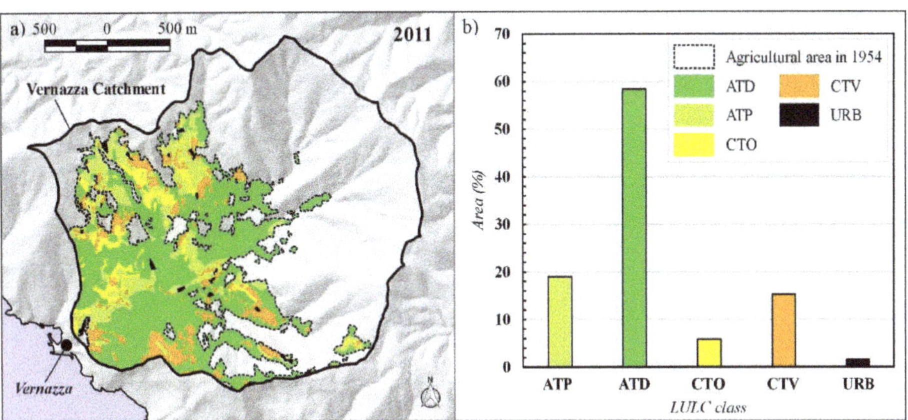

Figure 5. Land use/land cover (LULC) setting of the 1954 agricultural area in 2011. (**a**) LULC map. (**b**) Area in percentage of each LULC class; ATD, abandoned terraced slope with dense cover; ATP, abandoned terraced slope with poor cover; CTO, cultivated olive grove; CTV, cultivated vineyard; URB, urban area.

4.2. Relationship between Rainfall-Induced Shallow Landslides and Farmland Abandonment

After the disastrous rainstorm occurred on 25 October 2011, 364 shallow landslides were mapped by Cevasco et al. [78] through visual analysis of aerial orthophotographs and geomorphological surveys.

Overall, GIS analysis revealed that 295 of these rainfall-induced landslides have been triggered over agricultural terraced areas dated 1954 while nine landslides run-out through them from neighboring zones. The smallest slope failure covered a surface of a few square meters, the largest one covered 5376 m^2 while the average landslide size was estimated to be 216 m^2. The corresponding total area affected by landslides is 6.48 × 10^{-2} km^2, representing 2.75% and 1.12% of AGR 1954 and of the entire Vernazza catchment, respectively. Considering that the extent of 1954 cultivated slopes was about 2.35 km^2, the average landslide density in this land use class is 125 landslides/km^2. Generally, in 2011 the highest amount of rainfall-induced instability phenomena (68.8%) occurred on abandoned agricultural lands, which are the most represented land cover class (77.4%), while 31.2% of slope instabilities affected still cultivated terraced slopes (21.1%). In particular, as can be seen in Figure 6, most of the landslide crowns (39%) affected ATD while a lower percentage (29.8%) of the landslides has been triggered on ATP, representing 58.4% and 19.0% of 1954 agricultural areas, respectively. An important amount of landslide crowns was also detected in CTV (23.7%), while considerably lower landslide scars affected CTO (7.5%), which represent 15.3% and 5.8% of the areas that were cultivated in 1954, respectively. Furthermore, it was observed that landslides triggered on cultivated areas are generally lower in magnitude than those occurred on abandoned ones. In detail, the average extent of landslides that occurred on CT (70 m^2) is about half of those that occurred on ATP (140 m^2) and ATD (145 m^2). The most important observations come from the analysis of the landslide index, which represents the landslide area in each land use class in relation to the area of the considered land use class. The highest LI value was obtained for ATP (5.5%). Significant LI values were also associated to CTV (2.85%) and CTO (2.67%) whereas it was significantly lower for ATD (1.85%).

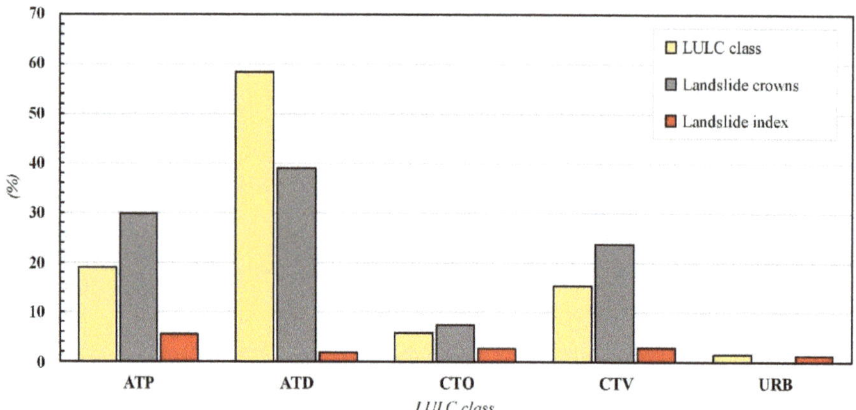

Figure 6. Histograms showing the relationships between the percentage of landslide crowns and of the Landslide Index with respect to land use/land cover of 2011 (LULC 2011); ATP, abandoned terraced slopes with poor cover; ATD, abandoned terraced slopes with dense cover; CTO cultivated olive groves; CTV, cultivated vineyards; URB, urban areas.

5. Discussion

The most important landscape transformations experienced by the Cinque Terre territory are directly or indirectly related to human action. Due to an impressive work of deforestation, reworking of eluvial-colluvial deposits, and building of dry-stone masonries, which has been carried out since the early centuries of the Middle Ages, terraced slopes have become the leading morphological peculiarity of the local landscape. During this time frame, farmland terraces have played a crucial role both as a source of livelihoods and as a regulating factor of hydro-geomorphological processes. According to Terranova et al. [80], the maximum extent of cultivated terraced slopes can be dated back to the second half of the 19th century. Starting from this period, terrace cultivation and maintenance have suffered from the first interruptions, which were mainly driven by scarce agricultural incomes that in turn led to the initiation of the earliest depopulation trends. However, the major decline of farmland started after the Second World War [80–82]. This period represents the so-called "Italian economic boom" that was the main cause of the largest exodus from agricultural lands [28]. On the other hand, the first decade of the new millennium can be assumed as the maximum expression of farmland abandonment of hilly and mountainous zones. Therefore, in this study these two periods were selected since they can be considered of crucial relevance in analyzing the effect of farmland abandonment on the hydro-geomorphological dynamics of slopes.

The results of this research revealed that approximately 40% of the Vernazza catchment was occupied by cultivated, that is terraced, areas in the early 1950s, denoting that farming activities were still an important driving force of the local economy. However, it should be noted that the extent of such areas cannot be considered as the actual total surface of the basin characterized by terraced slopes. In fact, an unknown portion of the investigated basin could have been abandoned before the early 1950s, causing it to look like a natural and/or semi-natural area in 1954 due to the presence of a dense vegetation cover. This is confirmed by the results provided by Terranova et al. [80] obtained through field surveys and analysis of historical aerial photos older than early 1950s, estimating that 51% of the whole Vernazza municipality territory (extending about 12 km^2) could be considered as occupied by terraces, both cultivated and characterized by different levels of abandonment.

This study also showed a very high agreement between the outcomes of land use mapping based on the interpretation of aerial photos and those obtained through cadastral data, confirming that these data sources are well comparable from both a geometric and a thematic point of view. These evidences are consistent with numerous researches reported in literature about the usefulness of old cadastral maps as a reliable reference about past LULC settings [62,65]. However, we have highlighted the remarkable spatial accuracy characterizing the considered cadastral maps, which is mainly related to the high density of cadastral parcels. This is a peculiar feature of Ligurian cadastral maps which strongly depend on the high fragmentation of land ownerships among numerous smallholders together with the rugged morphology of this region. This aspect is particularly relevant in the Cinque Terre area [79], where cadastral parcels can be very small in size, sometimes hardly representable due to scale. In addition, the more detailed LULC information associated to cadastral data allowed to better refine the land use setting through the identification of the different types of crop. These aspects testify that old cadastral data may be very useful to address question on farmland abandonment in historical time. For example, such data could represent effective tools to recognize the location and the extent of terraces at a very detailed scale, also where they have been covered by dense vegetation or disrupted by geomorphological processes. In this regard, cadastral-based spatial information may be coupled with high-resolution remote sensing techniques such as light detection and ranging (LiDAR) [83].

The LULC change detection pointed out a loss of cultivated areas amounting to 77.4% over approximately sixty years. As a result, such land use changes severely influenced the effects of hydro-geomorphological phenomena in these areas, as evidenced by the rainfall-induced ground effects of the 25 October 2011 event. This research showed that 1954 agricultural areas were strongly affected by landslides both in terms of number and magnitude (Figure 7). As can be seen in Figure 7b, rainfall-induced landslides triggered on 1954 agricultural areas were more numerous (295—81%) with respect to those triggered outside of them (69—19%). Moreover, in terms of the area covered by landslides (Figure 7c), considering that the percentage of the overall basin area affected by slope failures was about 1.5% [11], 75.2% of the total landslide area is referred to 1954 agricultural areas. These observations testify that terraced areas are highly susceptible to slope instabilities in case of extreme rainfall. This is in agreement with the evidences reported by other researches performed both in the study area [11,59,84–90] and in other terraced environments around the world [58,91–93]. As reported in technical literature, several factors can influence the occurrence of slope instabilities in terraced systems [58]. Some of these factors are related to the geometrical features of terraced slopes (e.g., height of dry-stone walls and slope steepness) [23] while in other cases they are directly associated to the hydrological functions of terraces. For example, the development of groundwater flows in response to the major water infiltration can lead to the saturation of backfill soils that in turn can cause a dangerous increase of pore water pressure with negative consequences on soil strength [32,91–93]. Other researchers outlined the disfunctions of the drainage systems (e.g., channels and ditches) [58] or the disturbance of dry-stone walls due to root growth as an important landslide predisposing agent. As reported by Cevasco et al. [11], within the study area the mass movements may have been favored by the permeability contrast between backfill soil and underlying bedrock that in turn may have promoted the formation of perched groundwater tables and seepage processes parallel to slope or upward directed. In this stratigraphic layout, sliding surfaces are often set at the interface between the retained soil and the bedrock [11,57,93]. The higher landslide susceptibility of terraced areas can be also attributed to the degree of maintenance of dry-stone masonries [53]. In the study area, the lack of maintenance can affect abandoned areas and to some extent also cultivated terraces [11].

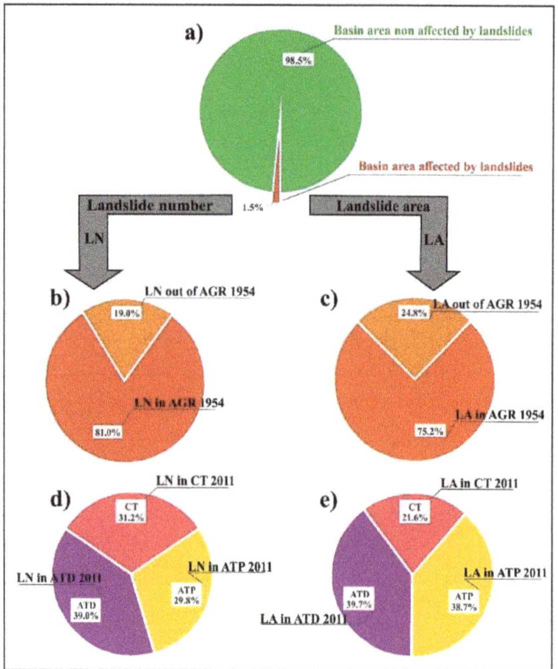

Figure 7. (**a**) Percentage of basin area affected by shallow landslides; (**b**) Percentage of landslides triggered in and out of AGR 1954; (**c**) Percentage of landslide area in and out of AGR 1954; (**d**) Percentage of landslides triggered in CT, ATP and ATD in 2011; (**e**) Percentage of landslide area in CT, ATP and ATD in 2011; LN, landslide number; LA, landslide area; AGR 1954, agricultural area of 1954; CT, cultivated terraced slopes; ATP, abandoned terraced slopes with poor cover; ATD, abandoned terraced slopes with dense cover.

This research clearly highlighted the negative role of farmland abandonment on rainfall-induced effects. The highest number of failure phenomena (68.8%) was triggered on abandoned agricultural lands (Figure 7d). However, the landslide index values reveal that ATP are characterized by higher landslide susceptibility than cultivated terraced slopes and ATD, respectively. Therefore, these outcomes further strengthen the findings of previous researches indicating that abandoned slopes become more stable when colonized by dense vegetation and thus after longer time of abandonment [59,94,95]. Conversely, since recently abandoned terraces are poorly colonized by forest tree species, they do not benefit from the stabilizing effects of vegetation (e.g., contribution of anchorage by root systems of forest tree species), so landslides processes are more intense [59,96–98]. Therefore, the phase between the initiation of terrace abandonment and the development of the dense vegetation represents the most hazardous scenario. Interestingly, the percentage of landslide area in CT is considerably lower than one-third of the total landslide area while it is approximately half of the landslide area in ATP and ATD. By comparing Figure 7d with Figure 7e it can be inferred that despite the number of triggered landslides being significant also in CT (31.2%, or 92 landslides), the corresponding landslide area (21.6%) is smaller than ATD (39.7%) and ATP (38.7%), respectively. In this regard, this research showed that shallow landslides triggered on CT were on average characterized by lower magnitude than those occurred on ATP and ATD. This outcome is consistent with similar studies [59] and testifies the negative consequences induced by the cessation of terrace cultivation and maintenance on slope stability. It should be noted, however, that in the study area the high fragmentation of land ownerships can make difficult the maintenance of dry-stone walls. In this regard, cultivated areas can be often placed near or surrounded by abandoned properties, making the maintenance works on border dry-stone walls not possible to be carried out (Figure 8).

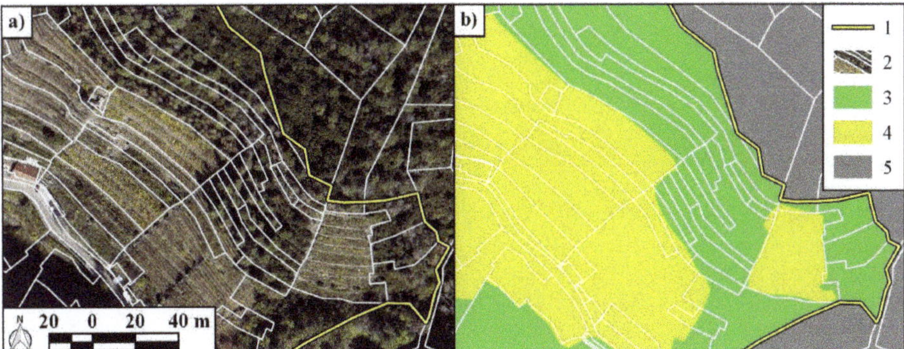

Figure 8. Comparison between the cadastral map and the 2011 orthophoto (**a**) and the 2011 LULC map (**b**). It can be noted the high fragmentation of land ownerships as well as the spatial accuracy of old cadastral data. 1, Boundary between the agricultural areas and the natural areas dated 1954; 2, parcel border; 3, abandoned terraced slope with dense cover (ATD); 4, cultivated vineyard (CTV); 5, natural and semi-natural area (NAT).

6. Conclusions

Farmland abandonment is a typical land use change commonly affecting hilly and mountainous zones worldwide. Terraced landscapes are extremely vulnerable territories, where crop abandonment can have extensive implications on the effects of hydrological and geomorphological processes, particularly in case of extreme rainfall events. In these peculiar landscapes, such effects represent a threat to slope stability and constitute an important source of risk when located in the proximity of urbanised areas.

This paper presented a quantitative multi-temporal analysis, performed in a GIS environment using different spatial information sources, aimed at investigating the LULC transformations that affected agricultural terraced slopes in a small coastal catchment and accurately estimating the degree of abandonment of terraces. In addition to this, the implications of such LULC modifications on the distribution, abundance, and magnitude of rainfall induced shallow landslides triggered by an intense rainstorm that occurred in 2011 were analysed.

It was observed that over the last sixty years extensive portions of agricultural terraced slopes have been abandoned and this has severely influenced the effects of shallow landsliding, producing negative consequences on slope stability. Since the selected study area is representative of farmland abandonment experienced by many rural territories of the entire Liguria region, our observations can support the scheduling of effective land management strategies. In this sense, we highlighted the usefulness of old cadastral data because of both the high spatial accuracy and thematic completeness of the associated database.

Eventually, the results of this research can be useful for future studies that address soil erosion and the shallow landslide susceptibility of terraced slopes by means of simulations of land use change scenarios.

Author Contributions: G.P. and A.M. have developed the conceptualization and the methodology of the study. A.M. performed GIS analysis. G.P. and A.M. have interpreted obtained results. G.P. written and managed the manuscript. E.R. and P.S. have contributed to data collection and discussion of results. A.C. and P.B. have supervised and coordinated the research activity and have provided their suggestions and revisions during the writing of the paper.

Funding: This research received no external funding.

Acknowledgments: The authors would like to thank Cinque Terre National Park for providing historical cadastral data of the study area. The authors wish to thank also three anonymous referees for their helpful comments and suggestions that improved this paper.

Conflicts of Interest: The authors declare no conflict of interest.

References

1. Van Westen, C.J.; Castellanos, E.; Kuriakose, S.L. Spatial data for landslide susceptibility, hazard, and vulnerability assessment: An overview. *Eng. Geol.* **2008**, *102*, 112–131. [CrossRef]
2. Corominas, J.; Van Westen, C.; Frattini, P.; Cascini, L.; Malet, J.P.; Fotopolou, S.; Catani, F.; Van Den Eeckhaut, M.; Mavrouli, O.; Agliardi, F.; et al. Recommendations for the quantitative analysis of landslide risk. *Bull. Eng. Geol. Environ.* **2014**, *73*, 209–263. [CrossRef]
3. Fohrer, N.; Haverkamp, S.; Eckhardt, K.; Frede, H.G. Hydrologic response to land use changes on the catchment scale. *Phys. Chem. Earth Part B* **2001**, *26*, 577–582. [CrossRef]
4. Gutierrez, F.; Gutierrez, M.; Sancho, C. Geomorphological and sedimentological analysis of a catastrophic flash flood in the Aràs drainage basin (central Pyrenees, Spain). *Geomorphology* **1998**, *22*, 265–283. [CrossRef]
5. Mandarino, A.; Maerker, M.; Firpo, M. The stolen space: A history of channelization, reduction of riverine areas and related management issues. The lower Scrivia River case study (NW Italy). *Int. J. Sustain. Dev. Plan.* **2019**, *14*, 118–129. [CrossRef]
6. Mandarino, A.; Maerker, M.; Firpo, M. Channel planform changes along the Scrivia River floodplain reach in northwest Italy from 1878 to 2016. *Quat. Res.* **2019**, *91*, 620–637. [CrossRef]
7. Glade, T. Landslide occurrence as a response to land use change: A review of evidence from New Zealand. *Catena* **2003**, *51*, 297–314. [CrossRef]
8. Tasser, E.; Mader, M.; Tappeiner, U. Effects of land use in alpine grasslands on the probability of landslides. *Basic Appl. Ecol.* **2003**, *4*, 271–280. [CrossRef]
9. Beguería, S. Changes in land cover and shallow landslide activity: A case study in the Spanish Pyrenees. *Geomorphology* **2006**, *74*, 196–206. [CrossRef]
10. Persichillo, M.G.; Bordoni, M.; Meisina, C. The role of land use changes in the distribution of shallow landslides. *Sci. Total Environ.* **2017**, *574*, 924–937. [CrossRef] [PubMed]
11. Cevasco, A.; Pepe, G.; Brandolini, P. The influences of geological and land use settings on shallow landslides triggered by an intense rainfall event in a coastal terraced environment. *Bull. Eng. Geol. Environ.* **2014**, *73*, 859–875. [CrossRef]
12. Galve, J.P.; Cevasco, A.; Brandolini, P.; Soldati, M. Assessment of shallow landslide risk mitigation measures based on land use planning through probabilistic modelling. *Landslides* **2015**, *12*, 101–114. [CrossRef]
13. Van Beek, L.P.H.; Van Asch, T.W. Regional assessment of the effects of land-use change on landslide hazard by means of physically based modelling. *Nat. Hazards* **2004**, *31*, 289–304. [CrossRef]
14. Gariano, S.L.; Petrucci, O.; Rianna, G.; Santini, M.; Guzzetti, F. Impacts of past and future land changes on landslides in southern Italy. *Reg. Environ. Chang.* **2018**, *18*, 437–449. [CrossRef]
15. Promper, C.; Puissant, A.; Malet, J.P.; Glade, T. Analysis of land cover changes in the past and the future as contribution to landslide risk scenarios. *Appl. Geogr.* **2014**, *53*, 11–19. [CrossRef]

16. Malek, Ž.; Boerboom, L.; Glade, T. Future forest cover change scenarios with implications for landslide risk: An example from Buzau Subcarpathians, Romania. *Environ. Manag.* **2015**, *56*, 1228–1243. [CrossRef]

17. Greenway, D.R. Vegetation and slope stability, chapter 6. In *Slope Stability*; Anderson, M.G., Richards, K.S., Eds.; John Wiley and Sons Ltd.: West Sussex, UK, 1987; pp. 187–230.

18. Wilkinson, P.L.; Anderson, M.G.; Lloyd, D.M. An integrated hydrological model for rain-induced landslide prediction. *Earth Surf. Proc. Landf.* **2002**, *27*, 1285–1297. [CrossRef]

19. Kosmas, C.; Danalatos, N.; Cammeraat, L.H.; Chabart, M.; Diamantopoulos, J.; Farand, R.; Gutiérrez, M.; Jacob, A.; Marques, H.; Martinez-Fernandez, J.; et al. The effect of land use on runoff and soil erosion rates under Mediterranean conditions. *Catena* **1997**, *29*, 45–59. [CrossRef]

20. Crozier, M.J. Multiple-occurrence regional landslide events in New Zealand: Hazard management issues. *Landslides* **2005**, *2*, 247–256. [CrossRef]

21. García-Ruiz, J.M. The effects of land uses on soil erosion in Spain: A review. *Catena* **2010**, *81*, 1–11. [CrossRef]

22. Del Monte, M.; Vergari, F.; Brandolini, P.; Capolongo, D.; Cevasco, A.; Ciccacci, S.; Conoscenti, C.; Fredi, P.; Melelli, L.; Rotigliano, E.; et al. Multi-method evaluation of denudation rates in small Mediterranean catchments. In *Engineering Geology for Society and Territory*; Lollino, G., Manconi, A., Clague, J., Shan, W., Chiarle, M., Eds.; Springer International Publishing: Cham, Switzerland, 2015; Volume 1, pp. 563–567. [CrossRef]

23. García-Ruiz, J.M.; Lana-Renault, N. Hydrological and erosive consequences of farmland abandonment in Europe, with special reference to the Mediterranean region—A review. *Agric. Ecosyst. Environ.* **2011**, *140*, 317–338. [CrossRef]

24. Lasanta, T.; Nadal-Romero, E.; Arnáez, J. Managing abandoned farmland to control the impact of re-vegetation on the environment. The state of the art in Europe. *Environ. Sci. Policy* **2015**, *52*, 99–109. [CrossRef]

25. Diodato, N.; Soriano, M.; Bellocchi, G.; Fiorillo, F.; Cevasco, A.; Revellino, P.; Guadagno, F.M. Historical evolution of slope instability in the Calore River Basin, Southern Italy. *Geomorphology* **2017**, *282*, 74–84. [CrossRef]

26. Poyatos, R.; Latron, J.; Llorens, P. Land use and land cover change after farmland abandonment. The case of a Mediterranean Mountain area (Catalan Pre-Pyrenees). *Mt. Res. Dev.* **2003**, *23*, 362–368. [CrossRef]

27. Vicente-Serrano, S.; Lasanta, T.; Romo, A. Analysis of the spatial and temporal evolution of vegetation cover in the Spanish Central Pyrenees. The role of human management. *Environ. Manag.* **2005**, *34*, 802–818. [CrossRef] [PubMed]

28. MacDonald, D.; Crabtree, J.R.; Wiesinger, G.; Dax, T.; Stamou, N.; Fleury, P.; Gutierrez Lazpita, J.; Gibon, A. Agricultural abandonment in mountain areas of Europe: Environmental consequences and policy response. *J. Environ. Manag.* **2000**, *59*, 47–69. [CrossRef]

29. Cerdà, A. Soil erosion after land abandonment in a semiarid environment of Southeastern Spain. *Arid Land Res. Manag.* **1997**, *11*, 163–176. [CrossRef]

30. Agnoletti, M. The degradation of traditional landscape in a mountain area of Tuscany during the 19th and 20th centuries: Implications for biodiversity and sustainable management. *For. Ecol. Manag.* **2007**, *249*, 5–17. [CrossRef]

31. Koulouri, M.; Giourga, C. Land abandonment and slope gradient as key factors of soil erosion in Mediterranean terraced lands. *Catena* **2007**, *69*, 274–281. [CrossRef]

32. Arnáez, J.; Lasanta, T.; Errea, M.P.; Ortigosa, L. Land abandonment, landscape evolution, and soil erosion in a Spanish Mediterranean mountain region: The case of Camero Viejo. *Land Degrad. Dev.* **2011**, *22*, 537–550. [CrossRef]

33. Brandolini, P.; Pepe, G.; Capolongo, D.; Cappadonia, C.; Cevasco, A.; Conoscenti, C.; Marsico, A.; Vergari, F.; Del Monte, M. Hillslope degradation in representative Italian areas: Just soil erosion risk or opportunity for development? *Land Degrad. Dev.* **2018**, *29*, 3050–3068. [CrossRef]

34. Gioia, D.; Lazzari, M. Testing the Prediction Ability of LEM-Derived Sedimentary Budget in an Upland Catchment of the Southern Apennines, Italy: A Source to Sink Approach. *Water* **2019**, *11*, 911. [CrossRef]

35. Cevasco, A.; Diodato, N.; Revellino, P.; Fiorillo, F.; Grelle, G.; Guadagno, F.M. Storminess and geo-hydrological events affecting small coastal basins in a terraced Mediterranean environment. *Sci. Total Environ.* **2015**, *532*, 208–219. [CrossRef]

36. Capolongo, D.; Diodato, N.; Mannaerts, C.; Piccarreta, M.; Strobl, R.O. Analyzing temporal changes in climate erosivity using a simplified rainfall erosivity model in Basilicata (southern Italy). *J. Hydrol.* **2008**, *356*, 119–130. [CrossRef]

37. Piccarreta, M.; Capolongo, D.; Boenzi, F.; Bentivenga, M. Implications of decadal changes in precipitation and land use policy to soil erosion in Basilicata, Italy. *Catena* **2006**, *65*, 138–151. [CrossRef]

38. Della Seta, M.; Del Monte, M.; Fredi, P.; Lupia Palmieri, E. Spacetime variability of denudation rates at the catchment and hillslope scales on the Tyrrhenian side of Central Italy. *Geomorphology* **2009**, *107*, 161–177. [CrossRef]

39. Piacentini, T.; Galli, A.; Marsala, V.; Miccadei, E. Analysis of soil erosion induced by heavy rainfall: A case study from the NE Abruzzo Hills Area in Central Italy. *Water* **2018**, *10*, 1314. [CrossRef]

40. Calista, M.; Miccadei, E.; Piacentini, T.; Sciarra, N. Morphostructural, Meteorological and Seismic Factors Controlling Landslides in Weak Rocks: The Case Studies of Castelnuovo and Ponzano (North East Abruzzo, Central Italy). *Geosciences* **2019**, *9*, 122. [CrossRef]

41. Carabella, C.; Miccadei, E.; Paglia, G.; Sciarra, N. Post-wildfire landslide hazard assessment: The case of the 2017 Montagna del Morrone fire (Central Apennines, Italy). *Geosciences* **2019**, *9*, 175. [CrossRef]

42. Pepe, G.; Mandarino, A.; Raso, E.; Cevasco, A.; Firpo, M.; Casagli, N. Extreme flood and landslides triggered in the Arroscia Valley (Liguria Region, Northwestern Italy) during the November 2016 rainfall event. In *Slope Stability: Case Histories, Landslide Mapping, Emerging Technologies, Proceedings of the IAEG/AEG Annual Meeting Proceedings, San Francisco, CA, USA, 17–21 September 2018*; Shakoor, A., Kato, K., Eds.; Springer International Publishing: Cham, Switzerland, 2019; Volume 1, pp. 171–175. [CrossRef]

43. Galve, J.P.; Cevasco, A.; Brandolini, P.; Piacentini, D.; Azañon, J.M.; Notti, D.; Soldati, M. Cost-based analysis of mitigation measures for shallow-landslide risk reduction strategies. *Eng. Geol.* **2016**, *213*, 142–157. [CrossRef]

44. Cevasco, A.; Pepe, G.; D'Amato Avanzi, G.; Giannecchini, R. Preliminary analysis of the November 10, 2014 rainstorm and related landslides in the lower Lavagna valley (eastern Liguria). *Ital. J. Eng. Geol. Environ.* **2017**, *1*, 5–15. [CrossRef]

45. Morgan, R.P.C. *Soil Erosion and Conservation*, 2nd ed.; Longman Group: Harlow, UK, 2015.

46. Stanchi, S.; Freppaz, M.; Agnelli, A.; Reinsch, T.; Zanini, E. Properties, best management practices and conservation of terraced soils in Southern Europe (from Mediterranean areas to the Alps): A review. *Quat. Int.* **2012**, *265*, 90–100. [CrossRef]

47. Li, X.H.; Yang, J.; Zhao, C.Y.; Wang, B. Runoff and sediment from orchard terraces in southeastern China. *Land Degrad. Dev.* **2014**, *25*, 184–192. [CrossRef]

48. Grove, A.T.; Rackham, O. *The Nature of Mediterranean Europe: An Ecological History*; Yale University Press: New Haven, CT, USA, 2003.

49. Cammeraat, L.H. Scale dependent thresholds in hydrological and erosion response of a semi-arid catchment in southeast Spain. *Agric. Ecosyst. Environ.* **2004**, *104*, 317–332. [CrossRef]

50. Gallart, F.; Llorens, P.; Latron, J.; Regüés, D. Hydrological processes and their seasonal controls in a smallMediterranean mountain catchment in the Pyrenees. *Hydrol. Earth Syst. Sci.* **2002**, *6*, 527–537. [CrossRef]

51. Tarolli, P.; Preti, F.; Romano, N. Terraced landscapes: From an old best practice to a potential hazard for soil degradation due to land abandonment. *Anthropocene* **2014**, *6*, 10–25. [CrossRef]

52. Gallart, F.; Llorens, P.; Latron, J. Studying the role of old agricultural terraces on runoff generation in a small Mediterranean mountainous basin. *J. Hydrol.* **1994**, *159*, 291–303. [CrossRef]

53. Camera, C.; Masetti, M.; Apuani, T. Rainfall, infiltration, and groundwater flow in a terraced slope of Valtellina (Northern Italy): Field data and modelling. *Environ. Earth Sci.* **2012**, *65*, 1191–1202. [CrossRef]

54. Lasanta, T.; Arnáez, J.; Oserin, M.; Ortigosa, L.M. Marginal lands and erosion in terraced fields in the Mediterranean mountains. A case study in the Camero Viejo (Northwestern Iberian System, Spain). *Mt. Res. Dev.* **2001**, *21*, 69–76. [CrossRef]

55. Brunori, E.; Salvati, L.; Antogiovanni, A.; Biasi, R. Worrying about 'Vertical Landscapes': Terraced Olive Groves and Ecosystem Services in Marginal Land in Central Italy. *Sustainability* **2018**, *10*, 1164. [CrossRef]

56. Lasanta, T.; Vicente-Serrano, S.M.; Cuadrat-Prats, J.M. Mountain Mediterranean landscape evolution caused by the abandonment of traditional primary activities: A study of the Spanish Central Pyrenees. *Appl. Geogr.* **2005**, *25*, 47–65. [CrossRef]

57. Lesschen, J.P.; Cammeraat, L.H.; Nieman, T. Erosion and terrace failure due to agricultural land abandonment in a semi-arid environment. *Earth Surf. Proc. Landf.* **2008**, *33*, 1574–1584. [CrossRef]

58. Moreno-de-las-Heras, M.; Lindenberger, F.; Latron, J.; Lana-Renault, N.; Llorens, P.; Arnáez, J.; Romero-Díaz, A.; Gallart, F. Hydro-geomorphological consequences of the abandonment of agricultural terraces in the Mediterranean region: Key controlling factors and landscape stability patterns. *Geomorphology* **2019**, *333*, 73–91. [CrossRef]

59. Brandolini, P.; Cevasco, A.; Capolongo, D.; Pepe, G.; Lovergine, F.; Del Monte, M. Response of terraced slopes to a very intense rainfall event and relationships with land abandonment: A case study from Cinque Terre (Italy). *Land Degrad. Dev.* **2018**, *29*, 630–642. [CrossRef]

60. Lambin, E.F. Modelling and monitoring land-cover change processes in tropical regions. *Prog. Phys. Geogr.* **1997**, *21*, 375–393. [CrossRef]

61. Parcerisas, L.; Marull, J.; Pino, J.; Tello, E.; Coll, F.; Basnou, C. Land use changes, landscape ecology and their socioeconomic driving forces in the Spanish Mediterranean coast (El Maresme County, 1850–2005). *Environ. Sci. Policy* **2012**, *23*, 120–132. [CrossRef]

62. Hamre, L.N.; Domaas, S.T.; Austad, I.; Rydgren, K. Land-cover and structural changes in a western Norwegian cultural landscape since 1865, based on an old cadastral map and a field survey. *Landsc. Ecol.* **2007**, *22*, 1563–1574. [CrossRef]

63. Olarieta, J.R.; Rodriguez-Valle, F.L.; Tello, E. Preserving and destroying soils, transforming landscapes: Soils and land-use changes in the Valles County (Catalunya, Spain) 1853–2004. *Land Use Policy* **2008**, *25*, 474–484. [CrossRef]

64. Bender, O.; Boehmer, H.J.; Jens, D.; Schumacher, K.P. Using GIS to analyse long-term cultural landscape change in Southern Germany. *Landsc. Urban Plan.* **2005**, *70*, 111–125. [CrossRef]

65. Cousins, S.A. Analysis of land-cover transitions based on 17th and 18th century cadastral maps and aerial photographs. *Landsc. Ecol.* **2001**, *16*, 41–54. [CrossRef]

66. Falcucci, A.; Maiorano, L.; Boitani, L. Changes in land-use/land-cover patterns in Italy and their implications for biodiversity conservation. *Landsc. Ecol.* **2007**, *22*, 617–631. [CrossRef]

67. Raska, P.; Klimes, J.; Dubisar, J. Using local archive sources to reconstruct historical landslide occurrence in selected urban regions of the Czech Republic: Examples from regions with different historical development. *Land Degrad. Dev.* **2015**, *26*, 142–157. [CrossRef]

68. Brandolini, P.; Faccini, F.; Pescetto, C. I paesaggi terrazzati d'Italia. I terrazzamenti della Liguria: Un bene culturale e del paesaggio a rischio. *L'Universo* **2008**, *88*, 204–221.

69. Brancucci, G.; Paliaga, G. *The Hazard Assessment in a Terraced Landscape: The Liguria (Italy) Case Study in the Interreg III Alpter Project. Geohazards—Technical, Economical and Social Risk Evaluation*; Berkeley Electronics Press: Berkeley, CA, USA, 2007; pp. 227–234.

70. Cevasco, A.; Pepe, G.; Brandolini, P. Shallow landslides induced by heavy rainfall on terraced slopes: The case study of the October, 25th, 2011 event in the Vernazza catchment (Cinque Terre, NW Italy). *Rend. Online Soc. Geol. Ital.* **2012**, *21*, 384–386.

71. Regione Liguria. Geoportale Regione Liguria. Liguria Region: Genova, Italy. Available online: https://geoportal.regione.liguria.it (accessed on 26 July 2019).

72. Giammarino, S.; Giglia, G. Gli elementi strutturali della piega di La Spezia nel contesto geodinamico dell'Appennino Settentrionale. *Boll. Soc. Geol. Ital.* **1990**, *109*, 683–692.

73. Cevasco, A. I fenomeni di instabilità nell'evoluzione della costa alta delle Cinque Terre (Liguria Orientale). *Studi Costieri* **2007**, *13*, 93–109.

74. Brandolini, P. The outstanding terraced landscape of the Cinque Terre coastal slopes (eastern Liguria). In *Landforms and Landscapes of Italy*; Soldati, M., Marchetti, M., Eds.; Springer International Publishing: Cham, Switzerland, 2017; pp. 235–244.

75. Mastronuzzi, G.; Aringoli, D.; Aucelli, P.P.C.; Baldassarre, M.A.; Bellotti, P.; Bini, M.; Biolchi, S.; Bontempi, S.; Brandolini, P.; Chelli, A.; et al. Geomorphological map of the Italian coast: From a descriptive to a morphodynamic approach. *Geogr. Fis. Din. Quat.* **2017**, *40*, 161–196. [CrossRef]

76. Cevasco, A.; Pepe, G.; Brandolini, P. Geotechnical and stratigraphic aspects of shallow landslides at Cinque Terre (Liguria, Italy). *Rend. Online Soc. Geol. Ital.* **2013**, *24*, 52–54.

77. Terranova, R.; Zanzucchi, G.; Bernini, M.; Brandolini, P.; Campobasso, S.; Faccini, F.; Renzi, L.; Vescovi, P.; Zanzucchi, F. Geology, geomorphology and wines in the Cinque Terre National Park (Liguria, Italy) [Geologia, geomorfologia e vini nel Parco Nazionale delle Cinque Terre (Liguria, Italia)]. *Boll. Soc. Geol. Ital.* **2006**, *6*, 115–128.

78. Cevasco, A.; Brandolini, P.; Scopesi, C.; Rellini, I. Relationships between geo-hydrological processes induced by heavy rainfall and land-use: The case of 25 October 2011 in the Vernazza catchment (Cinque Terre, NW Italy). *J. Maps* **2013**, *9*, 289–298. [CrossRef]

79. Galanti, Y.; Barsanti, M.; Cevasco, A.; D'Amato Avanzi, G.; Giannecchini, R. Comparison of statistical methods and multi-time validation for the determination of the shallow landslide rainfall thresholds. *Landslides* **2018**, *15*, 937–952. [CrossRef]

80. Terranova, R.; Brandolini, P.; Spotorno, M.; Rota, M.; Montanari, C.; Galassi, D.; Nicchia, P.; Leale, S.; Bruzzo, R.; Renzi, L.; et al. *Patrimoni de Marjades a la Mediterrania Occidental. Una Proposta de Catalogaciò*; Commissiò Europea DGX, Programa Raphael: Palma Di Mallorca, Spain, 2002; p. 243.

81. Terranova, R. Il paesaggio costiero agrario terrazzato delle Cinque Terre in Liguria. *Studi Ric. Geogr.* **1989**, *12*, 1–58.

82. Agnoletti, M.; Errico, A.; Santoro, A.; Dani, A.; Preti, F. Terraced Landscapes and Hydrogeological Risk. Effects of Land Abandonment in Cinque Terre (Italy) during Severe Rainfall Events. *Sustainability* **2019**, *11*, 235. [CrossRef]

83. Godone, D.; Giordan, D.; Baldo, M. Rapid mapping application of vegetated terraces based on high resolution airborne LiDAR. *Geomat. Nat. Hazards Risk* **2018**, *9*, 970–985. [CrossRef]

84. Schilirò, L.; Cevasco, A.; Esposito, C.; Scarascia Mugnozza, G. Shallow landslide initiation on terraced slopes: Inferences from a physically-based approach. *Geomat. Nat. Hazards Risk* **2018**, *9*, 295–324. [CrossRef]

85. Schilirò, L.; Cevasco, A.; Esposito, C.; Scarascia Mugnozza, G. Role of Land Use in Landslide Initiation on Terraced Slopes: Inferences from Numerical Modelling. In *Advancing Culture of Living with Landslides, Diversity of Landslide Forms, Workshop on World Landslide Forum, Ljubljana, Slovenia, 29 May–2 June 2017*; Mikoš, M., Casagli, N., Yin, Y., Sassa, K., Eds.; Springer: Cham, Switzerland, 2017; Volume 4, pp. 315–320. ISBN1 978-3-319-53484-8. ISBN2 978-3-319-53485-5. [CrossRef]

86. Zingaro, M.; Refice, A.; Giachetta, E.; D'Addabbo, A.; Lovergine, F.; De Pasquale, V.; Pepe, G.; Brandolini, P.; Cevasco, A.; Capolongo, D. Sediment mobility and connectivity in a catchment: A new mapping approach. *Sci. Total Environ.* **2019**, *672*, 763–775. [CrossRef]

87. Persichillo, M.G.; Bordoni, M.; Meisina, C.; Bartelletti, C.; Giannecchini, R.; D'Amato Avanzi, G.; Galanti, Y.; Barsanti, M.; Cevasco, A.; Brandolini, P.; et al. Remarks on the Role of Landslide Inventories in the Statistical Methods Used for the Landslide Susceptibility Assessment. In *Advancing Culture of Living with Landslides, Diversity of Landslide Forms, Workshop on World Landslide Forum, Ljubljana, Slovenia, 29 May–2 June 2017*; Mikoš, M., Tiwari, B., Yin, Y., Sassa, K., Eds.; Springer: Cham, Switzerland, 2017; pp. 759–766. [CrossRef]

88. Persichillo, M.G.; Bordoni, M.; Meisina, C.; Bartelletti, C.; Giannecchini, R.; D'Amato Avanzi, G.; Galanti, Y.; Cevasco, A.; Brandolini, P.; Galve, J.P.; et al. Shallow landslide susceptibility analysis in relation to land use scenarios. In *Landslides and Engineered Slopes. Experience, Theory and Practice*; Aversa, S., Cascini, L., Picarelli, L., Scavia, C., Eds.; Associazione Geotecnica Italiana: Rome, Italy, 2016; pp. 1605–1612, ISBN 978-1-138-02988-0. [CrossRef]

89. Bordoni, M.; Persichillo, M.G.; Meisina, C.; Cevasco, A.; Giannecchini, R.; D'Amato Avanzi, G.; Galanti, Y.; Bartelletti, C.; Brandolini, P.; Zizioli, D. Developing and testing a data-driven methodology for shallow landslide susceptibility assessment: Preliminary results. *Rend. Online Soc. Geol. Ital.* **2015**, *35*, 25–28. [CrossRef]

90. Raso, E.; Mandarino, A.; Pepe, G.; Di Martire, D.; Cevasco, A.; Calcaterra, D.; Firpo, M. Landslide Inventory of the Cinque Terre National Park, Italy. In *Slope Stability: Case Histories, Landslide Mapping, Emerging Technologies, Proceedings of the IAEG/AEG Annual Meeting Proceedings, San Francisco, CA, USA, 17–21 September 2018*; Shakoor, A., Kato, K., Eds.; Springer International Publishing: Cham, Switzerland, 2019; Volume 1, pp. 201–205. [CrossRef]

91. Arnáez, J.; Lana-Renault, N.; Lasanta, T.; Ruiz-Flaño, P.; Castroviejo, J. Effects of farming terraces on hydrological and geomorphological processes. A review. *Catena* **2015**, *128*, 122–134. [CrossRef]

92. Camera, C.; Apuani, T.; Masetti, M. Mechanisms of failure on terraced slopes: The Valtellina case (northern Italy). *Landslides* **2014**, *11*, 43–54. [CrossRef]

93. Crosta, G.B.; Dal Negro, P.; Frattini, P. Soil slips and debris flows on terraced slopes. *Nat. Hazard Earth Syst.* **2003**, *3*, 31–42. [CrossRef]

94. Cammeraat, E.; van Beek, R.; Kooijman, A. Vegetation succession and its consequences for slope stability in SE Spain. *Plant Soil* **2005**, *278*, 135–147. [CrossRef]

95. Ruecker, G.; Schad, P.; Alcubilla, M.M.; Ferrer, C. Natural regeneration of degraded soils and site changes on abandoned agricultural terraces in Mediterranean Spain. *Land Degrad. Dev.* **1998**, *9*, 179–188. [CrossRef]

96. López-Vicente, M.; Poesen, J.; Navas, A.; Gaspar, L. Predicting runoff and sediment connectivity and soil erosion by water for different land use scenarios in the Spanish Pre-Pyrenees. *Catena* **2013**, *102*, 62–73. [CrossRef]

97. Bazzoffi, P.; Gardin, L. Effectiveness of the GAEC standard of cross compliance retain terraces on soil erosion control. *Ital. J. Agron.* **2011**, *6*, 43–51. [CrossRef]

98. Camera, C.; Djuma, H.; Bruggeman, A.; Zoumides, C.; Eliades, M.; Charalambous, K.; Abate, D.; Faka, M. Quantifying the effectiveness of mountain terraces on soil erosion protection with sediment traps and dry-stone wall laser scans. *Catena* **2018**, *171*, 251–264. [CrossRef]

Article

Dynamics of Runoff and Soil Erosion on Abandoned Steep Vineyards in the Mosel Area, Germany

Manuel Seeger [1],*, Jesús Rodrigo-Comino [1,2], Thomas Iserloh [1], Christine Brings [1] and Johannes B. Ries [1]

[1] Physical Geography, Trier University, D-54286 Trier, Germany; rodrigo@uni-trier.de (J.R.-C.); iserloh@uni-trier.de (T.I.); c.brings@icp-geologen.de (C.B.); riesj@uni-trier.de (J.B.R.)
[2] Soil Erosion and Degradation Research Group, Department of Geography, University of Valencia, E-46010 Valencia, Spain
* Correspondence: seeger@uni-trier.de

Received: 24 September 2019; Accepted: 4 December 2019; Published: 9 December 2019

Abstract: The Mosel Wine region has suffered during the last decades a decrease in productive area, mostly on steep sloping vineyards. To avoid the spread of diseases, the extraction of grapevines on abandoned vineyards is mandatory in Rhineland-Palatinate. At the same time, the organic production of wine is growing slowly, but well established in the area. We assess in this paper the degree of the land-use changes, as well as their effect on runoff generation and sediment production, depending on the age of the abandonment, as well as the type and age of the land management, whether organic or conventional. Land use data were obtained to identify land-use change dynamics. For assessment of runoff generation and soil erosion, we applied rainfall simulation experiments on the different types of vineyard management. These were organically managed, conventionally managed and abandoned ones, all of varying ages. During the last decades of the last century, a decrease of around 30% of vineyard surface could be observed in Germany's Mosel Wine Region, affecting mostly the steep sloping vineyards. Despite a high variability within the types of vineyard management, the results show higher runoff generation, and soil erosion associated with recently installed or abandoned vineyards when compared to organic management of the vineyards, where erosion reached only 12%. In organic management, runoff and erosion are also reduced considerably, less than 16%, after a decade or more. Thus, organic vineyard management practices show to be very efficient for reduction of runoff and erosion. Consequently, we recommend to adopt as far as possible these soil management practices for sustainable land management of steep sloping vineyards. In addition, soil protection measures are highly recommended for vineyard abandonment according to the law.

Keywords: steep sloping vineyards; Mosel Wine Region; land abandonment; runoff; soil erosion; rainfall simulation experiments

---

1. Introduction

At least beginning with the Neolithic age, human impact has caused dramatic changes in landscape dynamics [1–3]. In Central Europe, and all over the globe, increased soil erosion has been directly linked to the intensification of agricultural land use at different stages of human culture [4–7]. Contrasting with a simplistic view on the geomorphological dynamics, model applications revealed at the landscape level that significant spatiotemporal interactions have to be considered to understand the overall sediment export [8], and thus to interpret landscape evolution [9].

In Mediterranean environments, García-Ruiz et al., as well as others, have identified extensification of agricultural practices as the reason for higher soil erosion rates [10–14]. Nevertheless, the runoff and sediment dynamics decreased substantially with increasing time after abandonment [11,15,16]. Later, Ries and others confirmed the high complexity of the processes, depending on the development or degradation status of the soils and landscape [17–21]. In Spain, land abandonment has been studied intensively in different landscapes of the Ebro Basin, Central Iberian Meseta, Southern and Eastern Peninsula. The studies confirmed that divergent dynamics of vegetation cover, soils and geomorphodynamics are affected by prior land use management practices such as the intensive use of herbicides, machinery, or amendments to increase the productivity [14,20–23]. Vineyard abandonment has been also in the focus of more recent research in the Mediterranean, emphasizing the effect of abandonment on terraced slopes [24–26]. Most of the authors highlight the necessity of land management measures after abandonment for reducing the negative effects due to soil erosion and terrace failure.

In contrast, there is only limited information on soil erosion intensity and rates based on measurements in Central Europe [27,28]. It is especially the Mosel region where the longest records on soil erosion in Germany are available [29–32]. During 25 years of monitoring by Gerold Richter in Mertesdorf in the Ruwer Valley, starting in the early 1970s, a scaled plot system showed that soil erosion relates mainly to the occurrence of heavy rainfall events. The average yearly erosion is quite low (~0.3 t ha^{-1} y^{-1}), but summer thunderstorms may cause erosion rates 10 times higher than the yearly average. The Mertesdorf experimental plot station has also contributed to setting the standards of soil protection in vineyards, proposing reduced tillage and intense mulching for soil protection. The data also gave clear hints on the highly negative effects of traditional deep ploughing on soil erodibility [32].

Experimental approaches are a common tool to understand geomorphodynamic surface processes, such as runoff generation and erosion at different scales and process intensities [33–36], to understand patterns of transport of solid and dissolved matter [37,38] and even for assessment of land use planning [39,40] and model parametrization [41]. Especially in Mediterranean and semi-arid landscapes, a vast collection of data has been generated with rainfall simulations on different surfaces and with different simulators (e.g., [35–42]). Unfortunately, in Central Europe, there has been little experimental research with rainfall simulators available to help understand runoff generation in the low mountain ranges of Eifel and Hunsrück in the western part of Germany. Hümann et al. [42] showed that compaction of soils due to heavy machinery such as harvesters was a major source of runoff within forest environments. Rainfall simulation experiments also showed the effect of water repellency on runoff generation: Butzen et al. [43] demonstrated that this effect is also present in temperate forests, mainly under coniferous trees, but also, depending on drought conditions, on soils under deciduous trees. In addition, Zemke [44] confirmed that forest roads have a high impact on runoff generation and routing in forest environments. Alpine runoff and erosion have also been recently in the focus of experimental research with rainfall simulation experiments [45–47]. Nevertheless, the results of rainfall simulation experiments have contributed substantially to our knowledge of fundamental soil physical processes [48–50], and of the effect of organic farming [51,52].

Land abandonment has complex interactions within the landscape that has been cultivated for centuries. On one hand, it may affect landscape dynamics on a large scale, such as reducing water resources [53], and on the other, changing the landscape's visual characteristics is a detriment for tourism [54]. The abandonment, or non-cultivation of vineyards, leads to very specific problems of dissemination of diseases such as black rot and phylloxera [55–57]. Thus, a consequent clearing is recommended [57] and has been mandatory in Rhineland-Palatinate since 28 November 1997 (*Drieschenverordnung*) for vineyards remaining unmanaged for more than 2 years. The clearing has to guarantee that no further growth is possible from the stumps. This is only possible by mechanically removing the root systems of the grapevines, which is labour intensive and expensive. In addition, some areas are too steep for the use of machinery. For the management of abandoned areas, especially the steep sloping ones, grazing with sheep and goats has been introduced. The objective is to suppress the massive proliferation of shrubs and the re-sprouting of grapevines.

With this paper, our purpose is to focus on the effects of vineyard management and abandonment on the intensities of runoff and soil erosion in one of the most emblematic viticultural areas in Western

Germany. We compare conventional and organic vineyard management practices with vineyard abandonment and examine the effect of time on runoff generation and soil erosion. In addition, as land abandonment has been investigated mainly as a Mediterranean problem, we want to glance at the dimension of the problem in the Mosel wine region.

2. Materials and Methods

2.1. Study Area

The Mosel viticultural region (Germany) comprises the Mosel Valley and the lower valley of the southern tributaries Saar and Ruwer (see Figure 1). In these valleys, the research sites can be found: Waldrach (Ruwer valley, UTM32 338144/5512279) and Kanzem (Saar valley, UTM32, 324616/5502957).

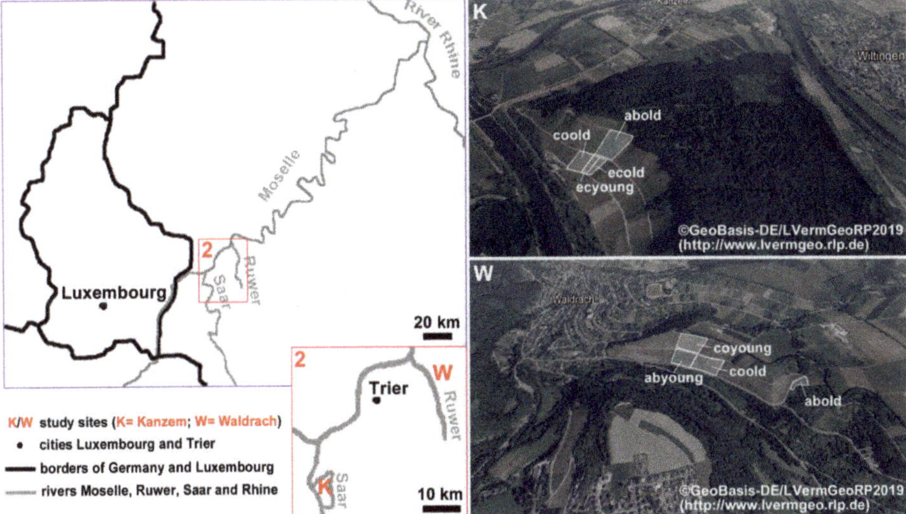

Figure 1. Location of the study sites Kanzem (**K**) and Waldrach (**W**). Management types: abold: abandoned old; abyoung: abandoned young; coold: conventional old; coyoung: conventional young; ecold: organic old; ecyoung: organic young.

The conventional old (coold) and young (coyoung) vineyards, and abandoned ones (abold and abyoung for the old and young, respectively) are situated in the village of Waldrach. They are cultivated in hill slopes with average elevations from 200 m to 400 m a.s.l. The main parent materials are Devonian greywackes, slates, and quartzites, which are in contact with Pleistocene fine materials transported by the Ruwer River, an affluent of the Moselle River [29]. Climate data is not available within the entire Ruwer Valley, but it is available close by (Station Avelsbach, 248 m a.s.l.). Within the last 10 years, the average temperature reached 10.2 °C, and the average annual rainfall was 732 mm.

According to IUSS-WRB-classification, soils are classified as Terric Anthrosols [58]. The active plantations are composed of 40-year-old (coold) and 5-year-old plants (coyoung, replanted 2012), and were cultivated at the summit of hillslopes. The old abandoned vineyard was cultivated since 1970 following a similar soil management system characterized by tillage, and was abandoned in 1990. Nowadays, this area is recolonized by spontaneous vegetation such as Rubus sp., Achillea millefolium agg. The young abandoned vineyard (abyoung) is located close to the conventional young one (coyoung). Grapevines were extracted in January 2008, and the slope was not managed any more for the following 4 years. Consequently, it was also recolonized with spontaneous pioneer herbaceous vegetation.

Conventional vineyard management keeps the herbaceous cover of the inter-rows in a changing mode every 2nd year. For weed control, the in that should be kept clear are harrowed at least once a year. Weed control directly below the grapevines takes place by mechanical weeding and the application of herbicides. For nutrient management, mineral fertilizer is applied, and also grape pomace is spread on the soils.

The plots are exposed SW-W and mean inclinations range from 15–25° although maximum values of 30° are reached in the backslope of the young conventionally managed vineyards. The annual total average rainfall is about 765 mm, and mean annual temperatures approximately 9 °C [59]. Soils are characterized by a thin organo-mineral horizon (nearly 2 cm deep) with high alteration induced by tillage and compaction (Ap). The horizon boundaries in Ap and B/C are abrupt (2–5 cm) and nearly plane in the compacted layers, and irregular in the rocky layer. However, in some parts, for example at the footslopes, one unique compacted and decapitate horizon can be found because of the depletion, trampling and tractor passes. In the old abandoned plot, soil profiles have a deep-rooted organic soil horizon (litter horizon O) from 0 to 4 cm, where no remains of any Ap horizon and large (>2 mm) soil aggregates are found. The boundaries with underlying soil horizons are also abrupt. Underneath, a B/C horizon characterized by tilled and compacted mineral soil can be observed with high rock fragment contents, and a weak soil structure characterized by prismatic and crumb forms of 20–50 mm size.

The other study area is within the municipality of Kanzem, at a vineyard location called Sonnenberg. The average annual temperature of the last 10 years is 11.3 °C, and the total annual rainfall 730 mm (Agrarmeteorologischer Dienst RLP—Agronomic-meteorological Service, www.am.rlp.de). The parent material, similar to that found in Waldrach, is based on Devonian slates, but at the toeslope it is intermixed with Pleistocene fluvial deposits, leading to soils with moderate to a high content of rock fragments. The soil texture is characterized by high silt content (40.4%). and organic matter (5.4%). They are classified as Terric Anthrosols according to the IUSS-WRB soil classification [58]

Here, between 165 and 195 m a.s.l., we can find next to each other an old, conventionally managed vineyard (coold), which was planted about 3 decades ago, and two organically managed vineyards, one young (ecyoung, 5 years) and one old (ecold, 55 years). All grapevines are variety Riesling. The abandoned vineyard (abold), located directly above the other ones and reaching an altitude of 230 m a.s.l., was removed from production in 1996. The slopes range between 20° and 35°.

The conventional plantation follows the same soil management system as the plots located in Waldrach. The organic vineyard management follows the standards set by the European Convention (regulation–EU-No 203/2012) and is certified yearly by independent surveyors. Tillage (harrowing, mulching and grubbing) between the grapevine rows, is mainly applied by machine from the end of March until May as well as in summer before the harvest. Nevertheless, a permanent cover with herbaceous species is kept throughout the year all over the vineyards. Hoeing for weed control under the rows and grass cutting is done by hand or with a brush cutter. The supply with nutrients is ensured by the application of compost, grape pomace, and horn shavings.

2.2. Data Sources and Methods

For understanding and quantifying the spatial and temporal dynamics of vineyard abandonment, we analysed the data gained and delivered by the Rhineland-Palatinate Statistical Office (Statistisches Landesamt Rheinland-Pfalz). The data is gathered in detail on all agricultural lands, also recording the type of grape cultivated. We used the data on the surface covered by vineyards (in ha) of the complete federal state Rhineland-Palatinate, for the Mosel viticultural region and the five districts within the region (Bernkastel, Obermosel, Burg Cochem, Ruwertal, Saar) on a yearly basis since 1999, and some data starting from 1964. In addition, we obtained data from the state agricultural agency (Landwirtschaftskammer Rheinland-Pfalz) for the years 1998, 2004, 2010, and 2015, indicating the surface area of vineyards on areas steeper than 30%. The recording on steep sloping vine growing surface has been stopped 2015, thus no newer data is available any more.

2.3. Rainfall Simulation Experiments

The dynamics of runoff and erosion by splash and shallow sheet flow were quantified by means of rainfall simulation experiments and conducted at different research periods (2008, 2015, and 2016). The small portable rainfall simulators (Figure 2) used as well as the standardized procedure have been described widely in [19,35,60]. The general pattern of the experiment was to apply a defined rainfall quantity, generated by pumping the water with a pump through a spraying nozzle (Lechler 460608), during 30 min on a small round plot (plot size 0.28 m^2, 60 cm diameter). The targeted intensity is 40 mm h^{-1}, which corresponds to approximately 5.7 L for each experiment. The runoff from the plot is collected completely in plastic bottles, subdividing the experiment into six different 5-min intervals from the beginning of the rainfall. Rainfall is controlled manually by a valve and a flow meter. The calibration of the rainfall intensity is done before and after the experiment to ensure a constant simulation quality collecting the complete runoff of the plot.

Figure 2. Rainfall simulation experiment: Pump is submerged in a blue water reservoir. Flow control by flow-meter and manometer. The image in the upper left corner shows circular plot. (Images by J. Rodrigo-Comino).

The amount of runoff and sediments is determined gravimetrically. The discharged water is filtered with Whatman filters No. 595 (pore size ~4 μm), and the filters weighted after drying at 105 °C for at least 24 h. Data on runoff and erosion is generated in 5-min intervals for each simulation as total runoff (Q (L)) and suspended sediment yield (SY (g)). In this case, only the overall sum for the entire simulation event was used. In addition, we calculated the specific runoff (spQ (L m^{-2}), specific sediment yield (spSY (g m^{-2})), suspended sediment concentration (SSC (g L^{-1}), and the runoff coefficient (RC [–]) for each of the experiments.

We differentiated three management types: Conventional, organic and abandoned. In addition, we characterized them by age: Young and old ones. Conventional and organic young vineyards have been installed 5 years ago or less, whilst the young abandoned vineyards were treated according to the law (extraction of the grapevines) less than 1 year before our testing. The old management types were managed or abandoned at least 15 years ago.

In total, we performed 82 rainfall simulation experiments on active or abandoned vineyards at the two study sites (see Table 1), 37 in Waldrach and 45 in Kanzem. 36 simulations were conducted on abandoned vineyards (Kanzem 14, Waldrach 22), and 46 on cultivated ones (Kanzem 31, Waldrach 15). Vineyards with organic management were located only in the Kanzem site, where 20 rainfall simulation experiments were performed.

Table 1. Distribution of management types and amount of rainfall simulations in both research sites. Management types: abold: abandoned old; abyoung: abandoned young; coold: conventional old; coyoung: conventional young; ecold: organic old; ecyoung: organic young.

Site	Management Group	RS Experiments
Kanzem	abold	14
	coold	11
	ecold	11
	ecyoung	9
Waldrach	abold	13
	abyoung	9
	coold	9
	coyoung	6

2.4. Data Analysis

Statistical analysis of the data was performed with with IBM SPSS Statistics 26 package [61]. Using the management categories of abandoned, conventional and organic combined with the age categories "young" and "old", we generated six different groups, including both test sites, Waldrach and Kanzem. Here, a simple explorative analysis was processed. The similarity of both study areas (Waldrach and Kanzem) was tested by means of two-tailed *t*-test. For understanding the differences between the management and age groups, we applied a Kruskal–Wallis test.

3. Results

3.1. Evolution of Wine-Growing Surface

The spatial and temporal development of vine growing during the last six decades shows opposing trends within the federal state of Rhineland-Palatinate (Figure 3). After a period of the apparently constant increase until the beginning of the 1990s, reaching a maximum (67,414 ha), recorded in 1997. For the last 10 years, the area covered by vineyards has stabilized at around 64,000 ha. In the Mosel Wine Region, the vineyards covered at the end of the eighties (1989) 12,509 ha, but decreased within two decades to around 8700 ha, stabilizing during the last 10 years at that level. The subareas of the Mosel Wine Region, where the test sites can be found, have therefore had a general decrease of the surface devoted to wine production, but with varying dynamics: The Ruwer Valley has experienced, after reaching a maximum of 252 ha of vineyards in 1997, a reduction of around 20%. The Saar valley lost more than 30% of the vineyards between 1979 (1194 ha) and 2006 (723 ha) but has experienced a slight increase of productive surface in the last decade.

Although there is only little data available on the cultivation of steep sloping vineyards, the obtained data shows the substantial reduction of these surfaces during the last decades. More than 30% of the steep sloping vineyards were lost, reducing their proportion of the total surface in the Mosel Wine Region from around 45% to around 40%. Thus, steep sloping vineyards are more affected by land abandonment than other vineyard surfaces.

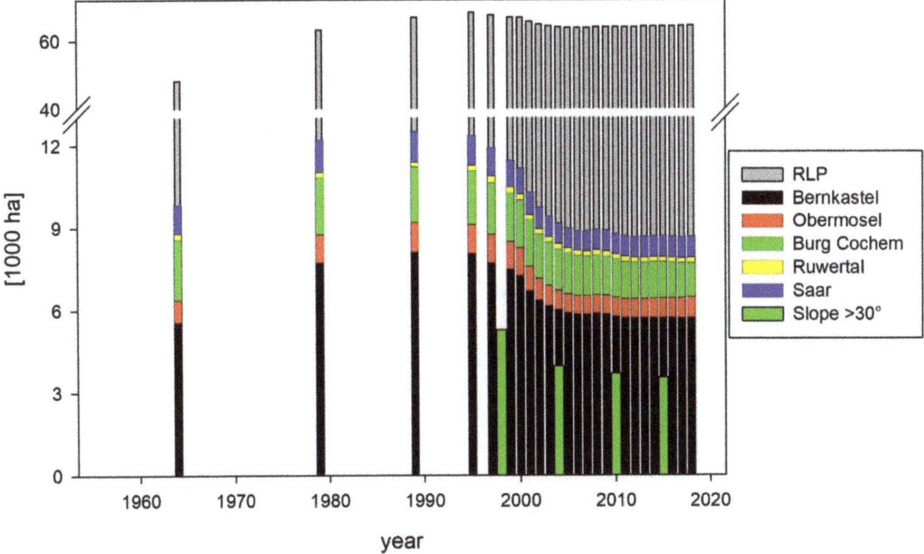

Figure 3. Surface covered by vineyard between 1964 and 2018. The total surface grown in Rhineland-Palatinate (RLP) is depicted, as well as the total area cultivated in the Mosel Wine Region, with details of all sub-sectors of the region. In addition, the surface of area cultivated on slopes >30° is given for 4 years between 1998 and 2015.

3.2. Surface Dynamics

The runoff dynamics of both areas, Kanzem and Waldrach, are quite similar (see Table 2), being higher in the latter area. Both areas show very high standard deviations on all parameters describing the initiation of runoff and erosion processes. This shows the high variability of the soils' key physical parameters because of cultivation due to spatially different intensities of compaction by wheel traffic, loosening by tillage, etc. In the Waldrach vineyards, maximum values for erosion parameters are substantially higher than in Kanzem. Nevertheless, the t-test shows a significant similarity between the two areas.

Table 2. Main descriptive values (avg: average; stddev: standard deviation; min: minimum; max: maximum) for the specific runoff (spQ (L m^{-2})), specific sediment yield (spSY (g m^{-2}), suspended sediment concentration (SSC (g L^{-1})) and runoff coefficient (RC (%)) for both test areas. The significance level of the *t*-test for each of the parameters is included.

| Variables | Area | | | | | | | | *t*-Test |
| | Kanzem | | | | Waldrach | | | | |
	avg	stddev	min	max	avg	stddev	min	max	Sig. (2-tailed)
spQ	2.3	3.9	0	16.9	3.4	5.3	0	25.9	0.29
RC	11.6	19.2	0	82.8	17.2	23.5	0	88.9	0.24
spSY	10.1	21.5	0	95.9	23.3	66.6	0	373.2	0.21
SSC	1.6	2.9	0	10.9	2.9	5.5	0	25.2	0.17

A look at the data subdivided by management type and age shows clear differences. Median specific runoff (see Table 3, Figure 4) is the highest in both the recently installed conventional vineyard (coyoung: 6.3 L m^{-2}) and the recently abandoned vineyard (abyoung: 6.1 L m^{-2}). In contrast, the lowest median runoff was recorded on the old organically managed vineyard (ecold: 0.0 L m^{-2}) and on the old abandoned one. In fact, more than half of the rainfall simulation experiments produced no runoff on these surfaces, and consequently no erosion. The average specific runoff is considerably higher on the old abandoned vineyards (abold: 1.2 L m^{-2}). In general, a higher specific runoff is found on established or recently abandoned vineyards. The variability of runoff generation is also higher on younger vineyards than on the older ones. It is striking that the young abandoned and conventionally managed vineyards show on all simulations some runoff (0.6 L m^{-2} and 0.2 L m^{-2}, respectively), while on all other areas some rainfall simulation experiments showed no runoff generation.

Table 3. Runoff parameters of the different management types. spQ: specific runoff (L m^{-2}); RC: runoff coefficient (%); des. par: descriptive parameters; avg: average; stdev: standard deviation; min: minimum; max: maximum. Management types: abold: abandoned old; abyoung: abandoned young; coold: conventional old; coyoung: conventional young; ecold: organic old; ecyoung: organic young.

des. par.	spQ						RC					
	abold	abyoung	coold	coyoung	ecold	ecyoung	abold	abyoung	coold	coyoung	ecold	ecyoung
avg	1.2	6.4	3.1	9.0	0.5	2.4	5.9	35.2	15.3	40.7	2.4	11.8
median	0.0	6.1	0.9	6.3	0.0	0.9	0.0	32.7	5.3	39.1	0.0	4.3
stdev	2.8	4.0	4.5	9.2	1.6	3.0	13.9	21.2	22.1	31.2	8.0	14.7
min	0.0	0.6	0.0	0.2	0.0	0.0	0.0	3.3	0.0	1.0	0.0	0.0
max	12.9	14.8	16.9	25.9	5.3	8.2	63.5	78.3	82.8	88.9	26.4	39.7

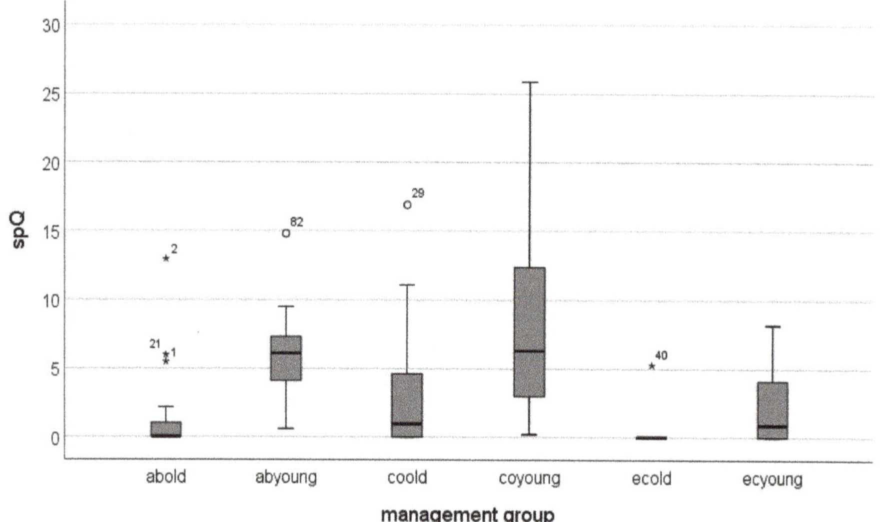

Figure 4. Boxplot of the specific runoff (L m^{-2}) of combined management and age classes. The bold line in the box shows the median, the box embraces the values between the 25% and 75% quantiles, the whiskers show data level 1.5× interquantile distance. Outliers are marked with "*" when outside 3× interquantile distance. Management groups are abold: abandoned old; abyoung: abandoned young; coold: conventional old; coyoung: conventional young; ecold: organic old; ecyoung: organic young.

Accordingly, the non-dimensional runoff coefficient shows the same trend (Table 3, Figure 5). But here, a look on the maximum values shows the high intensity of the processes: runoff coefficients around 80% are reached on all conventionally managed vineyards (coold: 83%; coyoung: 88.9%, as well as on the recently abandoned one (abyoung: 78.3%). The organic managed ones show maximum runoff coefficients clearly lower than 40% (ecold: 26.4%, ecyoung: 39.7%). On both organic vineyard groups, as well as the old abandoned vineyards, only a low or even no median runoff was generated (ecold: 0%; ecyoung: 4.3%; abold: 0%). Nevertheless, very high runoff can be also generated in the old abandoned area (max. abold: 63.5%).

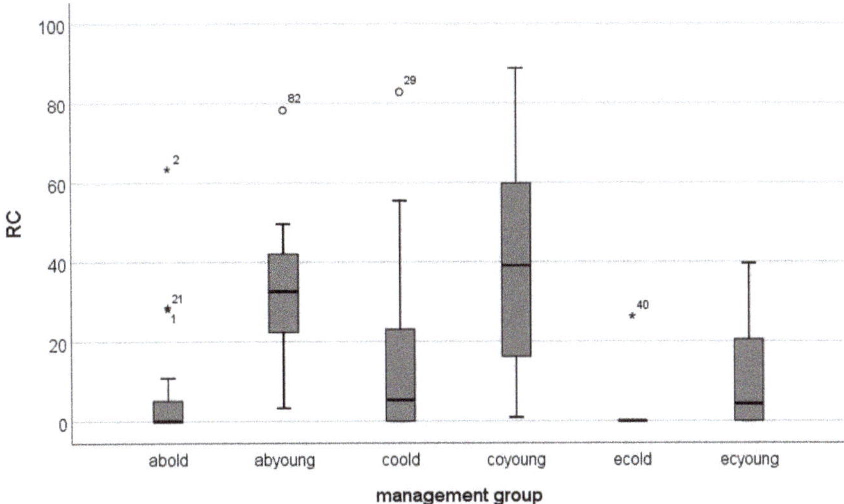

Figure 5. Boxplot of the runoff coefficient (%) of combined management and age classes. The bold line in the box shows the median, the box embraces the values between the 25% and 75% quantiles, the whiskers show data level 1.5× interquantile distance. Outliers are marked with "*" when outside 3× interquantile distance. Management groups are abold: abandoned old; abyoung: abandoned young; coold: conventional old; coyoung: conventional young; ecold: organic old; ecyoung: organic young.

Regarding the sediment yield, the data generated by rainfall simulation experiments show a much more complex situation (Table 4, Figure 6). The lowest average specific sediment yield is generated on old abandoned vineyards (abold: 1.6 g m^{-2}) and even lower on old organically managed ones (ecold: 0.5 g m^{-2}). According to the runoff generation (see above), there is no sediment yield on both management groups. However, the old abandoned areas show a substantially higher variability of sediment production by means of rainfall simulation experiments. Especially remarkable is the high amount of peak values, reaching up to 20.8 g m^{-2}. The vineyards managed for a longer period organically show up with a higher homogeneity in sediment production. Only one clear outlier is registered there, producing only 5.7 g m^{-2}.

Table 4. Descriptive statistics of the erosion parameters measured by means of rainfall simulation experiments. spSY: specific sediment yield (g m^{-2}); SSC: suspended sediment concentration (g l^{-1}); des. par: descriptive parameters; avg: average; stdev: standard deviation; min: minimum; max: maximum. Management types: abold: abandoned old; abyoung: abandoned young; coold: conventional old; coyoung: conventional young; ecold: organic old; ecyoung: organic young.

des. par.	spSY						SSC					
	abold	abyoung	coold	coyoung	ecold	ecyoung	abold	abyoung	coold	coyoung	ecold	ecyoung
Avg	1.6	64.7	17.8	38.7	0.5	10.3	0.3	6.5	3.2	4.7	0.1	2.1
median	0.0	10.0	1.3	7.6	0.0	2.7	0.0	2.7	2.1	2.6	0.0	1.3
stdev	4.7	121.5	28.1	52.7	1.7	18.3	0.8	8.5	3.7	6.6	0.3	2.6
min	0.0	1.8	0.0	0.6	0.0	0.0	0.0	0.8	0.0	0.7	0.0	0.0
max	20.8	373.2	95.9	119.3	5.7	56.7	3.3	25.2	10.9	18.0	1.1	6.9

The old vineyards managed conventionally, and the organic ones recently installed show median sediment yield values of 17.8 g m^{-2} and 10.3 g m^{-2}, respectively. Here, the conventionally managed vineyards show high variability, leading to a lower median (coold: 1.3 g m^{-2}; ecyoung: 2.7 g m^{-2}) and higher maximum sediment yield (coold: 95.9 g m^{-2}; ecyoung: 56.7 g m^{-2}).

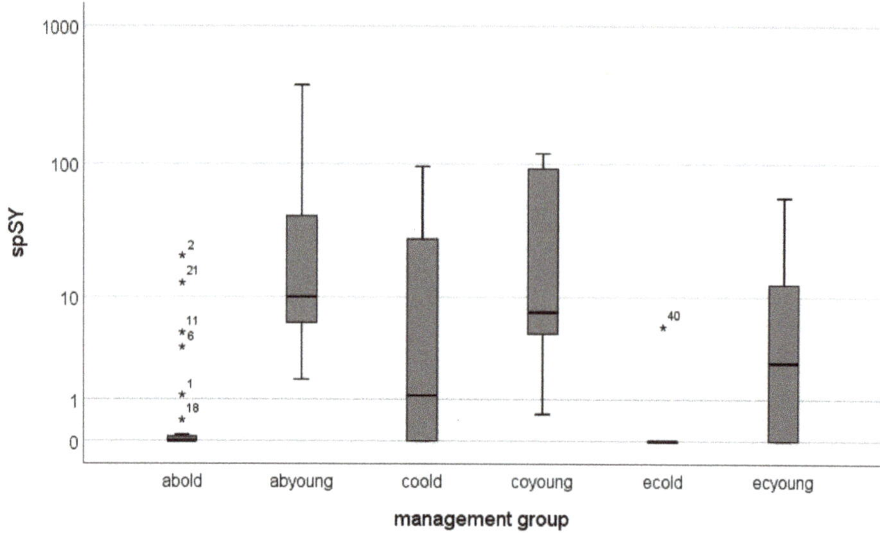

Figure 6. Boxplot of the specific sediment yield (g m^{-2}) of combined management and age classes. Mind the logarithmic y-axis! The bold line in the box shows the median, the box embraces the values between the 25% and 75% quantiles, the whiskers show data level 1.5× interquantile distance. Outliers are marked with "*" when outside 3× interquantile distance. Management groups are abold: abandoned old; abyoung: abandoned young; coold: conventional old; coyoung: conventional young; ecold: organic old; ecyoung: organic young.

The highest sediment yields, on average as well as extreme values, are measured on recently abandoned or installed conventional vineyards. With an average of 64.7 g m^{-2} (abyoung) and 38.7 g m^{-2} (coyoung) they are considerably higher than on all the other types of management. Also, the maximum values are substantially higher than on all the other groups (abyoung: 373.2 g m^{-2}; coyoung: 119.3 g m^{-2}).

Accordingly, the suspended sediment concentration paints a similar picture (Table 4, Figure 7). Old abandoned and organic vineyards are less prone to transport sediments in their runoff, with average values far below 1 g L^{-1} (abold: 0.34 g L^{-1}; ecold: 0.1 g L^{-1}). In addition, the measured maximum sediment concentrations are very low (abold: 3.3 g L^{-1}; ecold: 1.1 g L^{-1}). Older organic vineyards reach an intermediate level where average and maximum suspended sediment concentrations in the runoff reach 2.1 g L^{-1} and 6.9 g L^{-1}, respectively.

The situation changes drastically when regarding the remaining groups, as the average suspended sediment concentrations are very much higher than the ones mentioned above, ranging between 3.2 g L^{-1} (coold) and 6.5 g L^{-1} (abyoung). In addition, the maximum values are higher, reaching up to 25.2 g L^{-1} (abyoung). The variability of the measured values is the highest on the old conventional and young organic plots. However, on the young abandoned and conventional vineyards, low concentrations of transported sediments were measured, being them higher than the average on the old abandoned and organic ones.

The detection of similarities and dissimilarities by means of a Kruskal–Wallis test (see Table 5) shows clearly that only a few pairs of management types are similar over all runoff and sediment production patterns. These include the old abandoned and organically managed vineyards, the old abandoned and the young organically managed vineyards, and the vineyard with the old conventional management. Regarding the abandoned vineyards, which are in the focus of this article, the old abandoned ones are similar to the old and young organically managed ones, whilst the young abandoned ones are very similar to the recently installed conventional vineyards.

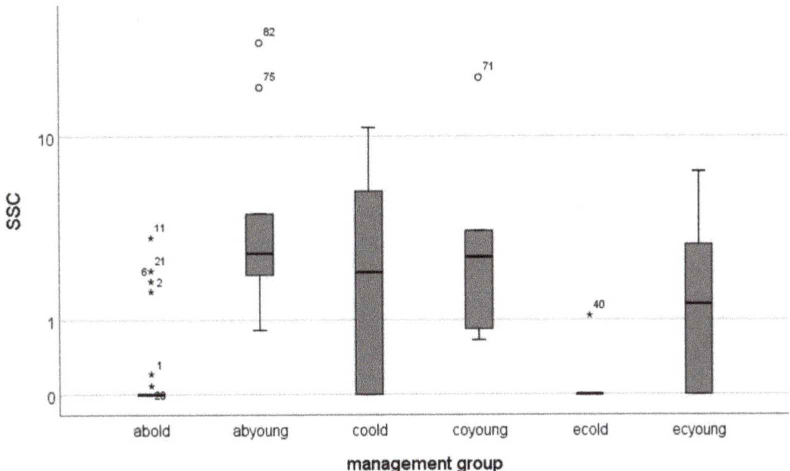

Figure 7. Boxplot of the suspended sediment concentration (g L^{-1}) of combined management and age classes. Mind the logarithmic y-axis. The bold line in the box shows the median, the box embraces the values between the 25% and 75% quantiles, the whiskers show data level 1.5× interquantile distance. Outliers are marked with "*" when outside 3× interquantile distance. Management groups are abold: abandoned old; abyoung: abandoned young; coold: conventional old; coyoung: conventional young; ecold: ecological old; ecyoung: ecological young.

Table 5. Results of the Kruskal–Wallis test for identifying pairwise similarities between the groups. Level of significance p is 0.05. All p-values < 0.05 are marked bold. Bonferroni-corrected significance is indicated by corr. p.

Group Pairs	spQ		RC		spSY		SSC	
	p	corr. p	p	corr. p	p	corr. p	p	corr. p
ecold-abold	0.382	1.000	0.380	1.000	0.457	1.000	0.572	1.000
ecold-ecyoung	0.071	1.000	0.078	1.000	0.046	0.968	0.035	0.731
ecold-coold	0.018	0.378	0.021	0.437	0.006	0.135	0.002	0.047
ecold-abyoung	0.000	0.001	0.000	0.001	0.000	0.010	0.001	0.020
ecold-coyoung	0.000	0.005	0.000	0.004	0.000	0.001	0.000	0.001
abold-ecyoung	0.195	1.000	0.215	1.000	0.102	1.000	0.052	1.000
abold-coold	0.051	1.000	0.061	1.000	0.010	0.217	0.001	0.029
abold-abyoung	0.000	0.001	0.000	0.001	0.001	0.018	0.001	0.023
abold-coyoung	0.001	0.012	0.000	0.009	0.000	0.001	0.000	0.001
ecyoung-coold	0.848	1.000	0.849	1.000	0.753	1.000	0.622	1.000
ecyoung-abyoung	0.026	0.552	0.020	0.419	0.097	1.000	0.167	1.000
ecyoung-coyoung	0.046	0.963	0.034	0.715	0.041	0.864	0.074	1.000
coold-abyoung	0.016	0.328	0.011	0.231	0.108	1.000	0.254	1.000
coold-coyoung	0.036	0.758	0.025	0.533	0.037	0.782	0.108	1.000
abyoung-coyoung	0.992	1.000	0.970	1.000	0.868	1.000	0.830	1.000

On the other hand, the old abandoned vineyards are dissimilar to all conventionally managed vineyards, and to the recently abandoned ones.

4. Discussion

The reduction of vineyards in the Mosel viticultural area, which contrasts with the generalized increase of surfaces planted with grapevines in the federal state (Rhineland-Palatinate), is generally related to structural and demographic conditions as well as the topography [62,63]. The latter makes vine-growing much cost and work-intensive, and thus prone to abandonment. The still generalized small size of the mostly family-owned vineyards and direct marketing may have also contributed to the trend. The substantial increase in prices in the last years [64] has not reversed this development but has at least seemed to stabilize the current situation. Despite the only limited data availability, there are with these several indications that most of the vineyard abandonment has occurred and would continue on steep slopes. The trend contrasts with the fact that the Mosel vineyards, located at the septentrional limit for wine-growing areas, are becoming in general beneficiaries of climate changes during the last decade and are expected to continue to do so [65,66]. The main reason stated by Levers et al. [67] for abandonment is, thus, not applicable here, but due to the socio-economic setting.

While data on the abandonment of steep slope vineyards is scarce, it is non-existent on the application of the mandatory rules (*Drieschenverordnung*) for vineyards' abandonment. Observations in the field show that there are several smaller vineyards not managed within the cultivated areas. For quantification and characterization of the types of abandonment, exhaustive research is needed. Nevertheless, the data on runoff and erosion gained here on abandoned vineyards represent surfaces treated according to the Federal State's regulation.

Most of the studies applying rainfall simulators systematically were conducted on Mediterranean areas, including agricultural and abandoned fields, vineyards and mountain regions [19,60,68–71], as well as in areas where continuous measurements have shown to be difficult or not accurate, or where rainfall variability is extremely high [14]. In Central Europe, the use or rainfall simulation experiments have been only used in few studies, which were the source of some of this data [59,72,73], or which focused on specific processes in forestry [43,44,74].

In general, runoff generation on all the vineyards investigated is in a similar order of magnitude as was recorded on stone-rich and heavily degraded soils in the Central Pyrenees of Spain [19,68], but higher than those recorded on vineyards in the Champagne [75]. The values are extremely high on the recently abandoned and installed conventional vineyards. The runoff generated on these surfaces is comparable to the runoff observed on crusted soils in the semi-arid central Ebro Basin [19] or southern Morocco [70]. Contrasting with this, the vineyards with organic management or those abandoned for a longer period show relatively low runoff generation, comparable with the ones recorded on less sloping vineyards in the French Champagne. This strong reduction of runoff generation after long-term abandonment has been also observed on other abandoned perennial and woody crops, which has been attributed to the increase of vegetation cover [76]. Nevertheless, the results from rainfall simulation experiments clearly contrast with the measurements of soil hydrological properties [77], as there were no clear differences detected between cultivated and managed vineyards in the Waldrach area. Thus, a much more complex interaction of soil properties and soil development has to be expected, depending on the management prior to the abandonment, as well as specific local situations, as has been stated in different studies [19,78].

Suspended sediment concentration values are quite high on the recently installed or abandoned vineyards, comparable to the values measured on Mediterranean vineyards, leading to extremely high total soil loss compared with values measured on vineyards in La Rioja (Spain) [79] or to other vineyards areas in the western Mediterranean Europe [80]. The measured values were even higher than those recorded on forest road cuts [81], but were predicted early, and identified as one of the major problems of soil erosion in vineyards [82]. On the other hand, the older abandoned, conventional and all organically managed vineyards show relatively low levels of suspended sediment concentration in their simulated runoff as well as total sediment yield, comparable to the abandoned fields in the Pyrenees or vineyards in Mediterranean areas [19,79,80]. Moreover, similar to there, we can observe on the older abandoned vineyards a high variability of erosion process intensities [68], with a predominance of very low trends to generate overland flow and erosion. In the case of the investigated vineyards, the reduction of erosion may be related to the depletion of fine sediments after abandonment. The resulting rock fragment cover has shown to be one of the most important factors controlling runoff generation and erosion [83]. In addition, the recolonization with vegetation is fast when soil and plant management stops. This leads also to a noteworthy reduction of runoff and sediment generation on older abandoned vineyards.

The very low geomorphodynamic activity of the organic vineyards is worth noticing. Here, the careful soil management, the systematic fostering of vegetation cover within the lanes, and the application of organic amendments, which have been proven to have fast positive effects on soil fertility and stability [84–87], show an effect that can be measured directly with experimental methods. These management strategies have been intensely discussed for decades in Germany [88–90], but long term studies, and our results, show that these measures are suitable for economically and organically sustainable viticulture [91].

5. Conclusions

A clear decrease of the surface of vineyards has occurred in the Mosel Wine Region, mainly during the last two decades of the 20th century and the first years of the 21st. As the reduction of the vineyards on steep slopes was over-proportional to the generalized decrease, it suggests that the difficult and expensive working conditions and the consequent lack of economic sustainability are the main triggers of abandonment. However, we had also to take into account a lack of basic data.

The results of rainfall simulation experiments show clear differences between the different investigated land-use systems. However, time of implementation of the vineyard management has proven to be an important factor modifying the effects of runoff and erosion in the vineyards. The organic vineyards are in general less prone to produce runoff, and also less susceptible to soil erosion than conventionally managed ones. Old organic vineyards are found to be the management systems with the lowest runoff and erosion detectable. This is mainly due to the fact that vineyards abandoned a long time ago may show on certain spots higher runoff and erosion patterns. In contrast, we can observe the differences in runoff and erosion dynamics for the recently installed vineyards. They produce high amounts of runoff, and their soils prove to be quite erodible, so the total amount of sediments produced are relatively high. Recently abandoned vineyards also show very high runoff and erosion when treated according to the law of Rhineland-Palatinate.

The results indicate clearly that suitable soil and vegetation management is fundamental for reducing water and soil losses on steep vineyards. In addition, strategies have to be developed for treating abandoned vineyards to avoid both spreading of disease, and water and soil loss.

Author Contributions: Conceptualization, M.S. and J.B.R.; methodology, M.S., J.R.-C., T.I., C.B.; formal analysis, M.S.; investigation, M.S., J.R.-C., C.B.; resources, M.S., J.B.R.; data curation, M.S., J.R.-C.; writing—original draft preparation, M.S.; writing—review and editing, M.S., J.R.-C., T.I.; supervision, M.S., J.B.R.; funding acquisition, M.S.

Funding: This research received no external funding. J.R.-C. was funded by Caixa-Bank and DAAD as well as the scholarship grant award from Ministerio de Educación, Cultura y Deporte de España (FPU15/01499).

Acknowledgments: The authors acknowledge the invaluable support for field and laboratory work by Bastienne Engels, Alexander Adrian, Cedric Röhrig, Martin Neumann and Rainer Bielen. We are also in debt to Weingut Steffes and Langguth (Waldrach) and Weingut Frei (Kanzem) for allowing the research on their vineyards.

Conflicts of Interest: The authors declare no conflict of interest.

References

1. Bork, H.-R.; Beckedahl, H.R.; Dahlke, C.; Geldmacher, K.; Mieth, A.; Li, Y. The world-wide explosion of soil erosion rates in the 20th century: The global soil erosion drama—Are we losing our food production base? *Petermanns Geographische Mitteilungen* **2003**, *147*, 16–29.

2. Bork, H.R. *Bodenerosion und Umwelt: Verlauf, Ursachen und Folgen der mittelalterlichen und neuzeitlichen Bodenerosion, Bodenerosionsprozesse, Modelle und Simulationen*; Landschaftsgenese und Landschaftsökologie; Abt. für Physische Geographie und Landschaftsökologie und für Physische Geographie und Hydrologie der Techn. University Selbstverl: Braunschweig, Germany, 1988.

3. Schmitt, A.; Dotterweich, M.; Schmidtchen, G.; Bork, H.-R. Vineyards, hopgardens and recent afforestation: Effects of late Holocene land use change on soil erosion in northern Bavaria, Germany. *Catena* **2003**, *51*, 241–254. [CrossRef]

4. Reiß, S.; Dreibrodt, S.; Lubos, C.C.M.; Bork, H.-R. Land use history and historical soil erosion at Albersdorf (northern Germany)—Ceased agricultural land use after the pre-historical period. *Catena* **2009**, *77*, 107–118. [CrossRef]

5.   Turkelboom, F.; Poesen, J.; Trébuil, G. The multiple land degradation effects caused by land-use intensification in tropical steeplands: A catchment study from northern Thailand. *Catena* **2008**, *75*, 102–116. [CrossRef]

6.   Peña, J.L.; Echeverria, M.T.; Julián, A.; Chueca, J. Processus d´accumulation et d´incision pendant l´Antiquité Classique dans la vallée de la Huerva (Bassin de l´Ebre, Espagne). In *Geoarchaeology of the Landscapes of Classical Antiquity*; Babesch Supplementa; Peeters: Leuven, Belgium, 2000; pp. 151–159. ISBN 978-90-429-0928-1.

7.   Sancho, C.; Gutierrez-Elorza, M.; Peña-Monné, J.L. Erosion and sedimentation during the Upper Holocene in the Ebro Depression: quantification and environmental significance. *Soil Eros. Stud. Spain* **1991**, 219–228.

8.   Bakker, M.M.; Govers, G.; van Doorn, A.; Quetier, F.; Chouvardas, D.; Rounsevell, M. The response of soil erosion and sediment export to land-use change in four areas of Europe: The importance of landscape pattern. *Geomorphology* **2008**, *98*, 213–226. [CrossRef]

9.   Lang, A.; Bork, H.-R.; Mäckel, R.; Preston, N.; Wunderlich, J.; Dikau, R. Changes in sediment flux and storage within a fluvial system: Some examples from the Rhine catchment. *Hydrol. Processes* **2003**, *17*, 3321–3334. [CrossRef]

10. Molinillo, M.; Lasanta, T.; García-Ruiz, J.M. Managing mountainous degraded landscapes after farmland abandonment in the Central Spanish Pyrenees. *Environ. Manag.* **1997**, *2*, 587–598. [CrossRef]

11. Garcia Ruiz, J.M.; Lasanta, T.; Ruiz-Flaño, P.; Martí, C.; Ortigosa, L.; Gonzáles, C. Soil erosion and desertification as a consequence of farmland abandonment in mountain areas. *Desertif. Control Bull.* **1994**, *25*, 27–33.

12. Garcia-Ruiz, J.M.; Lasanta, T.; Marti, C.; Gonzales, C.; White, S.; Ortigosa, L.; Ruiz Flano, P. Changes in runoff and erosion as a consequence of land-use changes in the Central Spanish Pyrenees. *Physics Chem. Earth* **1995**, *20*, 301. [CrossRef]

13. Llorens, P.; Queralt, L.; Plana, F.; Gallart, F. Studying solute and particulate sediment transfer in a small mediterranean mountainous catchment subject to land abandonment. *Earth Surf. Proc. Landf.* **1997**, *22*, 1027–1035. [CrossRef]

14. Cerdà, A.; Rodrigo-Comino, J.; Novara, A.; Brevik, E.C.; Vaezi, A.R.; Pulido, M.; Giménez-Morera, A.; Keesstra, S.D. Long-term impact of rainfed agricultural land abandonment on soil erosion in the Western Mediterranean basin. *Progress Phys. Geogr. Earth Environ.* **2018**, *42*, 202–219. [CrossRef]

15. Nadal-Romero, E.; Lasanta, T.; Regüés, D.; Lana-Renault, N.; Cerdà, A. Hydrological response and sediment production under different land cover in abandoned farmland fields in a mediterranean mountain environment. *Boletin de la Asociacion de Geografos Espanoles* **2011**, *55*, 303–323, 443–448.

16. Lasanta, T.; Perez-Rontome, C.; Garcia-Ruiz, J.M.; Machin, J.; Navas, A. Hydrological problems resulting from farmland abandonment in semi-arid environments: The central Ebro depression. *Physics Chem. Earth* **1995**, *20*, 309. [CrossRef]

17. Ries, J.B. Geomorphodynamics on fallow land and abandoned fields in the Ebro Basin and the Pyrenees—Monitoring of processes and development. *Z. Geomorph. N.F.* **2002**, *127*, 21–45.

18. Ries, J.B.; Marzolff, I.; Seeger, M. Influence of grazing on vegetation cover and geomorphodynamics. *Einfluss der Beweidung auf Vegetationsbedeckung und Geomorphodynamik Zwischen Ebrobecken und Pyrenäen* **2003**, *55*, 52–59.

19. Seeger, M. Uncertainty of factors determining runoff and erosion processes as quantified by rainfall simulations. *Catena* **2007**, *71*, 56–67. [CrossRef]

20. Romero-Díaz, A.; Ruiz-Sinoga, J.D.; Robledano-Aymerich, F.; Brevik, E.C.; Cerdà, A. Ecosystem responses to land abandonment in Western Mediterranean Mountains. *Catena* **2017**, *149*, 824–835. [CrossRef]

21. López-Vicente, M.; Nadal-Romero, E.; Cammeraat, E.L.H. Hydrological Connectivity Does Change Over 70 Years of Abandonment and Afforestation in the Spanish Pyrenees. *Land Degrad. Dev.* **2017**, *28*, 1298–1310. [CrossRef]

22. Ruiz-Sinoga, J.D.; Gabarrón Galeote, M.A.; Martinez Murillo, J.F.; Garcia Marín, R. Vegetation strategies for soil water consumption along a pluviometric gradient in southern Spain. *Catena* **2011**, *84*, 12–20. [CrossRef]

23. Bienes, R.; Marques, M.J.; Sastre, B.; García-Díaz, A.; Ruiz-Colmenero, M. Eleven years after shrub revegetation in semiarid eroded soils. Influence in soil properties. *Geoderma* **2016**, *273*, 106–114. [CrossRef]

24. Tarolli, P.; Preti, F.; Romano, N. Terraced landscapes: From an old best practice to a potential hazard for soil degradation due to land abandonment. *Anthropocene* **2014**, *6*, 10–25. [CrossRef]

25. Brandolini, P.; Cevasco, A.; Capolongo, D.; Pepe, G.; Lovergine, F.; Del Monte, M. Response of Terraced Slopes to a Very Intense Rainfall Event and Relationships with Land Abandonment: A Case Study from Cinque Terre (Italy). *Land Degrad. Dev.* **2018**, *29*, 630–642. [CrossRef]

26. Dunjó, G.; Pardini, G.; Gispert, M. Land use change effects on abandoned terraced soils in a Mediterranean catchment, NE Spain. *Catena* **2003**, *52*, 23–37. [CrossRef]

27. Auerswald, K.; Fiener, P.; Dikau, R. Rates of sheet and rill erosion in Germany—A meta-analysis. *Geomorphology* **2009**, *111*, 182–193. [CrossRef]

28. Cerdan, O.; Govers, G.; Le Bissonnais, Y.; Van Oost, K.; Poesen, J.; Saby, N.; Gobin, A.; Vacca, A.; Quinton, J.; Auerswald, K.; et al. Rates and spatial variations of soil erosion in Europe: A study based on erosion plot data. *Geomorphology* **2010**, *122*, 167–177. [CrossRef]

29. Richter, G. *Bodenerosion in Rebanlagen des Moselgebietes. Ergebnisse Quantitativer Untersuchungen 1974–1977*; Forschungsstelle Bodenerosion d. University Trier: Mertesdorf, Germany, 1979; Volume 3, p. 62.

30. Richter, G. On the soil erosion problem in the temperate humid area of Central Europe. *GeoJournal* **1980**, *4*, 279–287. [CrossRef]

31. Hacisalihoglu, S. Determination of soil erosion in a steep hill slope with different land-use types: a case study in Mertesdorf (Ruwertal/Germany). *J. Environ. Biol.* **2007**, *28*, 433–438.

32. Stehling, E.; Schmidt, R.G. *Das Datenarchiv der Forschungsstelle Bodenerosion in Mertesdorf (Ruwertal)*; Forschungsstelle Bodenerosion d. University Trier: Trier, Germany, 2017; Volume 16, p. 95.

33. Bowyer-Bower, T.A.S.; Burt, T.P. Rainfall simulators for investigating soil response to rainfall. *Soil Technol.* **1989**, *2*, 1–16. [CrossRef]

34. Bowyer-Bower, T.A.S.; Bryan, R.B. Rill initiation: concepts and experimental evaluation on badland slopes. *Zeitschrift fur Geomorphologie, Supplementband* **1986**, *60*, 161–175.

35. Iserloh, T.; Fister, W.; Seeger, M.; Willger, H.; Ries, J.B. A small portable rainfall simulator for reproducible experiments on soil erosion. *Soil Tillage Res.* **2012**, *124*, 131–137. [CrossRef]

36. Seeger, M. Experiments as tools in geomorphology. *Cuadernos de Investigación Geográfica* **2017**, *43*, 11. [CrossRef]

37. Jenkins, M.B.; Truman, C.C.; Siragusa, G.; Line, E.; Bailey, J.S.; Frye, J.; Endale, D.M.; Franklin, D.H.; Schomberg, H.H.; Fisher, D.S.; et al. Rainfall and tillage effects on transport of fecal bacteria and sex hormones 17β-estradiol and testosterone from broiler litter applications to a Georgia Piedmont Ultisol. *Sci. Total Environ.* **2008**, *403*, 154–163. [CrossRef] [PubMed]

38. Coles, A.E.; Wetzel, C.E.; Martínez-Carreras, N.; Ector, L.; McDonnell, J.J.; Frentress, J.; Klaus, J.; Hoffmann, L.; Pfister, L. Diatoms as a tracer of hydrological connectivity: are they supply limited? *Ecohydrology* **2016**, *9*, 631–645. [CrossRef]

39. Loch, R.J. Using rainfall simulation to guide planning and management of rehabilitated areas: Part I. Experimental methods and results from a study at the northparkes mine, Australia. *Land Degrad. Dev.* **2000**, *11*, 221–240. [CrossRef]

40. Loch, R.J.; Connolly, R.D.; Littleboy, M. Using rainfall simulation to guide planning and management of rehabilitated areas: Part II. computer simulations using parameters from rainfall simulation. *Land Degrad. Dev.* **2000**, *11*, 241–255. [CrossRef]

41. Schindewolf, M.; Schmidt, J. Parameterization of the EROSION 2D/3D soil erosion model using a small-scale rainfall simulator and upstream runoff simulation. *Catena* **2012**, *91*, 47–55. [CrossRef]

42. Hümann, M.; Schüler, G.; Müller, C.; Schneider, R.; Johst, M.; Caspari, T. Identification of runoff processes – The impact of different forest types and soil properties on runoff formation and floods. *J. Hydrol.* **2011**, *409*, 637–649. [CrossRef]

43. Butzen, V.; Seeger, M.; Marruedo, A.; de Jonge, L.; Wengel, R.; Ries, J.B.; Casper, M.C. Water repellency under coniferous and deciduous forest—Experimental assessment and impact on overland flow. *Catena* **2015**, *133*, 255–265. [CrossRef]

44. Zemke, J.J. Runoff and Soil Erosion Assessment on Forest Roads Using a Small Scale Rainfall Simulator. *Hydrology* **2016**, *3*, 25. [CrossRef]

45. Kaiser, A.; Neugirg, F.; Schindewolf, M.; Haas, F.; Schmidt, J. Simulation of rainfall effects on sediment transport on steep slopes in an Alpine catchment. *Proc. Int. Assoc. Hydrol. Sci.* **2015**, *367*, 43–50. [CrossRef]

46. Hänsel, P.; Schindewolf, M.; Eltner, A.; Kaiser, A.; Schmidt, J. Feasibility of High-Resolution Soil Erosion Measurements by Means of Rainfall Simulations and SfM Photogrammetry. *Hydrology* **2016**, *3*, 38. [CrossRef]

47. Schindler Wildhaber, Y.; Bänninger, D.; Burri, K.; Alewell, C. Evaluation and application of a portable rainfall simulator on subalpine grassland. *Catena* **2012**, *91*, 56–62. [CrossRef]

48. Van den Putte, A.; Govers, G.; Leys, A.; Langhans, C.; Clymans, W.; Diels, J. Estimating the parameters of the Green–Ampt infiltration equation from rainfall simulation data: Why simpler is better. *J. Hydrol.* **2013**, *476*, 332–344. [CrossRef]

49. Langhans, C.; Govers, G.; Diels, J.; Clymans, W.; Van den Putte, A. Dependence of effective hydraulic conductivity on rainfall intensity: Loamy agricultural soils. *Hydrol. Proc.* **2010**, *24*, 2257–2268. [CrossRef]

50. Langhans, C.; Govers, G.; Diels, J.; Leys, A.; Clymans, W.; Putte, A.V.D.; Valckx, J. Experimental rainfall-runoff data: Reconsidering the concept of infiltration capacity. *J. Hydrol.* **2011**, *399*, 255–262. [CrossRef]

51. Ulrich, U.; Zeiger, M.; Fohrer, N. Soil structure and herbicide transport on soil surfaces during intermittent artificial rainfall. *Zeitschrift für Geomorphologie, Supplementary Issues* **2013**, *57*, 135–155. [CrossRef]

52. Zeiger, M.; Fohrer, N. Impact of organic farming systems on runoff formation processes—A long-term sequential rainfall experiment. *Soil Tillage Res.* **2009**, *102*, 45–54. [CrossRef]

53. Beguería, S.; López Moreno, J.I.; Lorente, A.; Seeger, M.; Garcia Ruiz, J.M. Assessing the Effect of Climate Oscillations and Land-use Changes on Streamflow in the Central Spanish Pyrenees. *Ambio* **2003**, *32*, 283–286. [CrossRef]

54. Job, H.; Murphy, A. Germany's Mosel Valley: Can Tourism Help Preserve Its Cultural Heritage? *Tour. Rev. Int.* **2006**, *9*, 333–347. [CrossRef]

55. Forneck, A.; Eder, J.; Schmid, J. Reblaus - Neue Gefahr: Blattgallen. *Der Deutsche Weinbau* **2017**, 34–37.

56. Molitor, D.; Baus, O.; Berkelmann-Löhnertz, B. Schwarzfäule—Was gibt es Neues? *Das Deutsche Weinmagazin* **2010**, *26*, 31–34.

57. Redl, H. Verwilderte Weingärten und Stilllegungsflächen mit hohem Gefahrenpotenzial. *Der Winzer* **2006**, *62*, 13–17.

58. IUSS Working Group WRB. *World Reference Base for Soil Resources 2014, Update 2015 International Soil Classification System for Naming Soils and Creating Legends for Soil Maps*; World Soil Resources Reports; FAO: Rome, Italy, 2015; ISBN 978-92-5-108370-3.

59. Rodrigo Comino, J.; Brings, C.; Lassu, T.; Iserloh, T.; Senciales, J.M.; Martínez Murillo, J.F.; Ruiz Sinoga, J.D.; Seeger, M.; Ries, J.B. Rainfall and human activity impacts on soil losses and rill erosion in vineyards (Ruwer Valley, Germany). *Solid Earth* **2015**, *6*, 823–837. [CrossRef]

60. Ries, J.B.; Marzen, M.; Iserloh, T.; Fister, W. Soil erosion in Mediterranean landscapes–Experimental investigation on crusted surfaces by means of the Portable Wind and Rainfall Simulator. *J. Arid Environ.* **2014**, *100–101*, 42–51. [CrossRef]

61. IBM Corp. *IBM SPSS 26 Statistics for Windows*; IBM Corp.: Armonk, NJ, USA, 2019.

62. Loose, S.; Strub, L. Winzergenossenschaften: Bewirtschaftung von Steillagen. *Wein + Markt* **2017**, *28*, 24–26.

63. Loose, S.; Pabst, E. Strukturelle Unterschiede: Sebstvermarkter. *Der Deutsche Weinbau* **2018**, *23*, 27–31.

64. Gutzler, M. Preissteigerungen im Weingut umsetzen: Analyse von Preislisten. *Das Deutsche Weinmagazin* **2016**, *20*, 19–21.

65. Jones, G.V.; White, M.A.; Cooper, O.R.; Storchmann, K. Climate Change and Global Wine Quality. *Clim. Change* **2005**, *73*, 319–343. [CrossRef]

66. Ashenfelter, O.; Storchmann, K. Using Hedonic Models of Solar Radiation and Weather to Assess the Economic Effect of Climate Change: The Case of Mosel Valley Vineyards. *Rev. Econom. Stat.* **2010**, *92*, 333–349. [CrossRef]

67. Levers, C.; Schneider, M.; Prishchepov, A.V.; Estel, S.; Kuemmerle, T. Spatial variation in determinants of agricultural land abandonment in Europe. *Sci. Total Environ.* **2018**, *644*, 95–111. [CrossRef]

68. Seeger, M.; Ries, J.B. Soil degradation and soil surface process intensities on abandoned fields in Mediterranean mountain environments. *Land Degrad. Dev.* **2008**, *19*, 488–501. [CrossRef]

69. Cerdà, A. The effect of season and parent material on water erosion on highly eroded soils in eastern Spain. *J. Arid Environ.* **2002**, *52*, 319–337. [CrossRef]

70. Peter, K.D.; d'Oleire-Oltmanns, S.; Ries, J.B.; Marzolff, I.; Ait Hssaine, A. Soil erosion in gully catchments affected by land-levelling measures in the Souss Basin, Morocco, analysed by rainfall simulation and UAV remote sensing data. *Catena* **2014**, *113*, 24–40. [CrossRef]

71. Snelder, D.J.; Bryan, R.B. The use of rainfall simulation tests to assess the influence of vegetation density on soil loss on degraded rangelands in the Baringo District, Kenya. *Catena* **1995**, *25*, 105–116. [CrossRef]

72. Kirchhoff, M.; Rodrigo-Comino, J.; Seeger, M.; Ries, J.B. Soil erosion in sloping vineyards under conventional and organic land use managements (Saar-Mosel Valley, Germany). *Cuadernos de Investigación Geográfica* **2017**, *43*, 22. [CrossRef]

73. Rodrigo-Comino, J.; Brings, C.; Iserloh, T.; Casper, M.C.; Seeger, M.; Senciales, J.M.; Brevik, E.C.; Ruiz-Sinoga, J.D.; Ries, J.B. Temporal changes in soil water erosion on sloping vineyards in the Ruwer-Mosel Valley. The impact of age and plantation works in young and old vines. *J. Hydrol. Hydromech.* **2017**, *65*, 402–409. [CrossRef]

74. Butzen, V.; Seeger, M.; Wirtz, S.; Huemann, M.; Mueller, C.; Casper, M.; Ries, J.B. Quantification of Hortonian overland flow generation and soil erosion in a Central European low mountain range using rainfall experiments. *Catena* **2014**, *113*, 202–212. [CrossRef]

75. Morvan, X.; Naisse, C.; Malam Issa, O.; Desprats, J.F.; Combaud, A.; Cerdan, O. Effect of ground-cover type on surface runoff and subsequent soil erosion in Champagne vineyards in France. *Soil Manag.* **2014**, *30*, 372–381. [CrossRef]

76. Cerdà, A.; Ackermann, O.; Terol, E.; Rodrigo-Comino, J. Impact of Farmland Abandonment on Water Resources and Soil Conservation in Citrus Plantations in Eastern Spain. *Water* **2019**, *11*, 824. [CrossRef]

77. Comino, J.R.; Bogunovic, I.; Mohajerani, H.; Pereira, P.; Cerdà, A.; Ruiz Sinoga, J.D.; Ries, J.B. The Impact of Vineyard Abandonment on Soil Properties and Hydrological Processes. *Vadose Zone J.* **2017**, *16*. [CrossRef]

78. Novák, T.J.; Incze, J.; Spohn, M.; Glina, B.; Giani, L. Soil and vegetation transformation in abandoned vineyards of the Tokaj Nagy-Hill, Hungary. *Catena* **2014**, *123*, 88–98. [CrossRef]

79. Arnaez, J.; Lasanta, T.; Ruiz-Flaño, P.; Ortigosa, L. Factors affecting runoff and erosion under simulated rainfall in Mediterranean vineyards. *Soil Tillage Res.* **2007**, *93*, 324–334. [CrossRef]

80. Rodrigo Comino, J.; Iserloh, T.; Morvan, X.; Malam Issa, O.; Naisse, C.; Keesstra, S.D.; Cerdà, A.; Prosdocimi, M.; Arnáez, J.; Lasanta, T. Soil erosion processes in European vineyards: a qualitative comparison of rainfall simulation measurements in Germany, Spain and France. *Hydrology* **2016**, *3*, 6. [CrossRef]

81. Arnáez, J.; Larrea, V.; Ortigosa, L. Surface runoff and soil erosion on unpaved forest roads from rainfall simulation tests in northeastern Spain. *Catena* **2004**, *57*, 1–14. [CrossRef]

82. Auerswald, K.; Schwab, A. Erosion risk (C factor) of different viticultural practices. *Die Wein-Wissenschaft Viticult. Enolog. Sci.* **1999**, *54*, 54–60.

83. Rodrigo Comino, J.; Iserloh, T.; Lassu, T.; Cerdà, A.; Keestra, S.D.; Prosdocimi, M.; Brings, C.; Marzen, M.; Ramos, M.C.; Senciales, J.M. Quantitative comparison of initial soil erosion processes and runoff generation in Spanish and German vineyards. *Sci. Total Environ.* **2016**. [CrossRef]

84. Novara, A.; Gristina, L.; Saladino, S.S.; Santoro, A.; Cerdà, A. Soil erosion assessment on tillage and alternative soil managements in a Sicilian vineyard. *Soil Tillage Res.* **2011**, *117*, 140–147. [CrossRef]

85. Prosdocimi, M.; Jordán, A.; Tarolli, P.; Keesstra, S.; Novara, A.; Cerdà, A. The immediate effectiveness of barley straw mulch in reducing soil erodibility and surface runoff generation in Mediterranean vineyards. *Sci. Total Environ.* **2016**, *547*, 323–330. [CrossRef]

86. Schwertmann, U.; Rickson, R.J.; Auerswald, K. Soil erosion protection measures in Europe. In Proceedings of the European Community workshop, Freising, Germany, 24–26 May 1988; Schweizerbart Science Publishers: Stuttgart, Germany, 1991.

87. Emde, K. Erosion exposure of the Hessen winegrowing regions, Germany. Recommendations for protection measures with view to climatic conditions. In *Proceedings of the Foerderungsdienst, Wien*; Bundesministerium für Land- und Forstwirtschaft; Wien: Vienna, Austria, 1994; Volume Sonderausgabe, pp. 45–51.

88. Kiefer, W. Cultivation of steeply sloping vineyards. *Rebe Wein* **1976**, *29*, 6–10.

89. Joerger, V.; Stoermann-Belting, J.; Zipf, S. Results for 1992 of the pilot project environment-friendly viticulture. *Der Badische Winzer* **1993**, *18*, 276–281.

90. Reimers, H.; Steinberg, B.; Kiefer, W. Root observations on grapevine dependent on different soil management systems. *Die Wein-Wissenschaft Viticult. Enolog. Sci.* **1994**, *49*, 136–145.

91. Götzke, A. *Entwicklung einer Naturschutzkonzeption in Weinbaugebieten auf der Grundlage einer Vergleichenden Untersuchung Faunistischer und Betriebswirtschaftlicher Parameter Praxisüblich und Ökologisch Erzeugender Weinbaubetriebe*; Bayerische Julius-Maximilians-Universität: Würzburg, Germany, 2006.

Article

Projecting Future Impacts of Global Change Including Fires on Soil Erosion to Anticipate Better Land Management in the Forests of NW Portugal

Amandine Valérie Pastor [1,2,*], Joao Pedro Nunes [2], Rossano Ciampalini [1], Myke Koopmans [3], Jantiene Baartman [3], Frédéric Huard [4], Tomas Calheiros [2], Yves Le-Bissonnais [1], Jan Jacob Keizer [5] and Damien Raclot [1]

[1]   LISAH, INRA, IRD, Montpellier SupAgro, University Montpellier, FR, 34060 Montpellier, France; rossano.ciampalini@supagro.fr (R.C.); yves.le-bissonnais@inra.fr (Y.L.-B.); damien.raclot@ird.fr (D.R.)
[2]   cE3c, Centre for Ecology, Evolution and Environmental Changes, Faculdade de Ciências, Universidade de Lisboa, 1749-016 Lisbon, Portugal; jpcnunes@fc.ul.pt (J.P.N.); tlmenezes@fc.ul.pt (T.C.)
[3]   Soil Physics and Land Management Group, Wageningen University, 6708 PB Wageningen, The Netherlands; myke.koopmans@wur.nl (M.K.); jantiene.baartman@wur.nl (J.B.)
[4]   INRA AgroClim, CEDEX 09, 84914 Avignon, France; frederic.huard@inra.fr
[5]   CESAM, Centre for Environmental and Marine Studies, Eco-Hydrology Laboratory & Earth Surface Processes Team, Department of Environment and Planning, University of Aveiro, 3810-193 Aveiro, Portugal; jjkeizer@ua.pt
*   Correspondence: amandine.pastor22@gmail.com

Received: 1 October 2019; Accepted: 4 December 2019; Published: 11 December 2019

Abstract: Wildfire is known to create the pre-conditions leading to accelerated soil erosion. Unfortunately, its occurrence is expected to increase with climate change. The objective of this study was to assess the impacts of fire on runoff and soil erosion in a context of global change, and to evaluate the effectiveness of mulching as a post-fire erosion mitigation measure. For this, the long-term soil erosion model LandSoil was calibrated for a Mediterranean catchment in north-central Portugal that burnt in 2011. LandSoil was then applied for a 20-year period to quantify the separate and combined hydrological and erosion impacts of fire frequency and of post-fire mulching using four plausible site-specific land use and management scenarios (S1. business-as-usual, S2. market-oriented, S3. environmental protection and S4. sustainable trade-off) and an intermediate climate change scenario Representative Concentration Pathway (RCP) 4.5 by 2050. The obtained results showed that: (i) fire had a reduced impact on runoff generation in the studied catchment (<5%) but a marked impact on sediment yield (SY) by about 30%; (ii) eucalypt intensification combined with climate change and fires can increase SY by threefold and (iii) post-fire mulching, combined with riparian vegetation maintenance/restoration and reduced tillage at the landscape level, was highly effective to mitigate soil erosion under global change and associated, increased fire frequency (up to 50% reduction). This study shows how field monitoring data can be combined with numerical erosion modeling to segregate the prominent processes occurring in post forest fire conditions and find the best management pathways to meet international goals on achieving land degradation neutrality (LDN).

Keywords: sediment yield; runoff; fire frequency; erosion control techniques; mulching; global change

---

1. Introduction

Erosion is a serious environmental problem worldwide including on-site effects (e.g., depletion in soil organic carbon stocks and decline in agronomic yields) and off-site effects (e.g., non-point source pollutions and reservoir siltation) [1]. Although it is a natural phenomenon, it can be exacerbated by climate change; fire and grazing [2] and is often accelerated by human activities [3]. The problem is especially disruptive in the Mediterranean area where erosion rates measured at the outlet of river basins of different sizes are high [4,5], especially because soils are usually thinner than in northern Europe [6] but also because croplands of the Mediterranean have a predominance of fallow land and lack of adoption of conservation practices [7]. The Mediterranean basin is very prone to erosion in all its forms because of the climate, topography, soil characteristics and a very long history of human presence and intense cultivation [8,9]. In addition, the occurrence of human-related forest fires affecting thousands of hectares each year is also a significant problem in both the northern and southern areas of the Mediterranean basin because wildfires have frequently been reported to produce strong and sometimes extreme hydrological and erosion responses in recently burnt areas, especially during the first few post-fire years [10,11].

Mediterranean-type ecosystems are particularly prone to fires, as the cool wet season is propitious for vegetation growth and fuel production, while the summer dry season promotes the drying of fuel and propitious conditions for fire occurrence and spread [12]. In the north-western Iberian Peninsula, the more humid climate has favored the commercial plantation of fast-growing but fire-prone species such as eucalypts and maritime pines, further exacerbating the wildfire problem [13]. Besides damaging vegetation and leading to loss of life and property, fires have a large number of indirect impacts, which may be more important in the long term, such as impacts on health through smoke inhalation; disruptions of social functions such as road and air traffic or business closures; environmental impacts in burned areas such as enhanced soil erosion, flooding and water contamination in and downstream of burned areas and long-term disruption of social-ecological dynamics [14]. In the Mediterranean, the enhancement of soil erosion in burned areas is particularly disruptive because, once the forest soils are directly exposed to the water action, they can be rapidly degraded by post-fire erosion and associated carbon and nutrient losses [15,16]. The export of fine sediments, associated with ashes and nutrients, can also contaminate streams and impact aquatic ecosystems and human water resources [17–19].

Fires are strongly dependent on climate [12,20]. Therefore, climate change projections of warmer and, in many fire-prone regions, drier climates indicate an increase of fire frequency and extend in the Mediterranean [21] and other regions [22,23]. However, fires are also dependent on vegetation, both in terms of potential vegetation growth and fuel production [12] and of changes to vegetation distribution, caused by climate or socioeconomic changes [24,25]. The resulting changes in wildfire regimes may lead to changes in their impacts, and requires an assessment of both the magnitude of these changes and the effectiveness of existing measures to mitigate them. The impacts of wildfire on soil fertility losses can often be effectively mitigated through soil conservation measures, especially when applied as emergency post-fire interventions. These measures include the application of organic residues to the soil surface (i.e., mulching, with straw or forest logging residues), log and shrub barriers, infiltration enhancement through scarifying or plowing or re-seeding and ecological restoration [17,26–29]. However, it is typically impractical to treat all burnt areas entirely, thus requiring knowledge of the effectiveness of these measures under different fire and post-fire conditions [29–34].

The impact of fire frequency on post-fire hydrological and erosion response has not yet been fully established for the forests of NW Iberia (Portugal and Spain) [35], and approaches to prioritize post-fire recovery are not widely applied in practice, except in Galicia [34]. More knowledge is also needed about the potential combined impacts of fires and global change on runoff and soil erosion, given the potential to exacerbate the already damaging impacts of fires on land degradation. The objective of this study was to assess the potential of soil conservation measures, including post-fire mulching, to mitigate the impacts of global change in a Mediterranean forested catchment of NW Portugal.

To this end, the long-term soil erosion model LandSoil was first calibrated using rainfall-runoff-erosion measurements on a catchment that burnt in 2011, and then applied over a 20-year period to quantify the individual and combined impacts on runoff and sediment yield (SY) of fire frequency (from no fire to two fires every 20 years), post-fire mulching and global change scenarios derived from four plausible land use and management scenarios under an intermediate climate change scenario.

2. Material and Methods

2.1. Study Site Macieira

The study was carried out in the Macieira de Alcôba experimental watershed, of 96 ha, located at an altitude of 470 m asl in the Caramulo mountain range of north-central Portugal (Figure 1). The climate can be classified as humid Mediterranean (Csb in the Koppen-Geiger classification) with annual precipitations ranging from 818 to 1294 mm (Pousadas rainfall station). About 70% of the precipitation falls during autumn and winter. The average air temperature in winter is 8 °C, while that in summer is 17 °C. Soils are shallow, mostly Luvisols and Cambisols, overlaying moderately impermeable bedrock of schist and granite [15,19]. The Southern part of the watershed contains the village surrounded by terraced agricultural fields with a rotation of maize in summer and pasture in winter, irrigated year-round with the "águas da lima" system [19]. Until the 1930s, the watershed was mainly occupied by agricultural fields on terraces and by shrublands on the steepest slopes; from the 1930s to 1970s, there was a large-scale afforestation of shrublands and part of the agricultural lands with Maritime Pine (*Pinus pinaster*); from the 1970s to present, afforestation with eucalypt (*Eucalyptus globulus*) occurred [36]. Around the village and agricultural area, the forest is present on slopes steeper than 5%. With a reliable map of burnt areas starting in the mid-1970s, major fires have affected the region about once per decade [37].

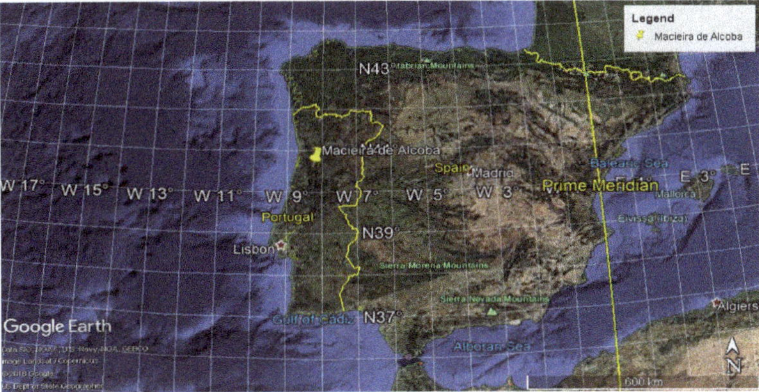

Figure 1. Location of Macieira de Alcoba, Portugal.

2.2. Data Collection

The Macieira watershed was monitored between 01 November 2010 and 30 September 2014, with continuous measurements of rainfall and other climatic variables, and streamflow and water turbidity at the outlet (Table 1). Observation showed that sediment exportation at the outlet is mainly due to suspended material because bedload is negligible in this site [28]. A fire burned c. 10% of the watershed on 11th August 2011, and about half of the burnt area was clear-cut and ploughed in February and March 2012; therefore, the available data set included two periods: before the fire (2010–2011) and after fire (2012–2014). Outlet data collection was complemented by a survey of linear erosion features composed of rills and ephemeral gullies after major storms, which the resulting volumes of eroded soil converted to mass using bulk density measurements. Furthermore, a characterization of soil texture, depth, hydrological properties (water retention curve, saturated hydraulic conductivity and others), roughness and shear strength, was carried out through a survey at 25 points within the catchment. A more detailed description of data collection can be found in Nunes et al. (2018) [38].

Table 1. Geomorpological and Hydrological Research Unit (HRU) characterization.

Characteristics	Value
Altitude	440 to 620 m asl
Gravelius compactness coefficient	1.2
Average Slope	16%
Average channel slope	11%
Drainage density	2.4 km km^{-2}
Landuse	forest (60%), cropland (23%), terraced cropland (12%), urban (5%)
Geology	granite (lowland), schist (slopes)
Soil	CMu (lowland), LPu (slopes)

2.3. Model Setup and Calibration Procedure

The LandSoil model (LandSoil: landscape design for soil conservation under land use and climate change) is a spatial raster-based model for simulating water and tillage erosion and landscape topography evolution over time spans on the orders of 10–100 years [39,40]. After simulating each event, LandSoil recalculates the elevation raster. Soil surface properties control water infiltration, runoff and sediment concentration for each grid-cell and rainfall event. As the result of water infiltration balance, runoff is routed over the catchment using a D8 algorithm [41] integrating the upstream/upslope cells contribution and the runoff generated within each cell. Linear landscape elements as ditches and tillage rows are also used to impose preferential directions to the flow [42].

LandSoil can simulate both rill and inter-rill erosion processes. The first is based on factors such as runoff friction and cohesion, topography (slope) and accumulated flow in an empirical relationship with the observed rill sections [43]. The latter, responsible for the remobilization of the soil particles detached by splash erosion, is controlled by the sediment concentration responding to rainfall intensity and soil surface properties such as vegetation cover, soil roughness and soil crusting as declined in expert rules issue of observations at parcel scale [44,45]. In LandSoil, sediment concentration is limited by threshold functions combining local topography, including profile curvature (concavity > 0.055 m^{-1}) and slope gradient (<0.02 mm^{-}), and land use/vegetation cover (>60%). Sediment concentration limits the range from 2.5 to 10 gl^{-}. Tillage erosion is simulated following the formulation of Govers et al. (1994) [46], both contour and downslope were modeled in different fields with tillage transport coefficients spanning from 111 to 139 kgm^{-g} [47].

The implementation of LandSoil on the studied catchment was based on a 2 m resolution grid based on a digital elevation model (DEM) derived from interpolating 1:10,000 contour lines using the ridge line method [48]. The set-up for LandSoil modeling mainly consists on defining the monthly calendars for soil roughness and crusting, vegetation cover, and tillage operation per land use based on a field sampling campaign in 2010 to determine soil texture, chemistry and origin (Table 1, Tables S4–S9). A digital elevation model (DEM) derived from interpolating 1:10,000 contour lines using the ridge line method [48], at 25 m of spatial resolution has been used to represent the catchment topography. In LandSoil, climate forcing is provided through a time series of precipitation events expressed through the effective volume and duration of precipitation, previous precipitation over 48 h and maximum precipitation over five minutes. The calibration of the model was made using events that occurred during the four years monitoring period in a 2-steps procedure: first for runoff, and then for erosion. Calibration for runoff consisted in manually adjusting the soil infiltration parameter. Calibration for erosion consisted in manually adjusting the sediment concentration in runoff and the rill sections. Parameters optimization was based on runoff and sediment exportation at the catchment outlet, as well as field measurements of hydrological soil properties and rill distribution sizes. In LandSoil, soil and land use parameters evolution are provided at a monthly base although simulations are made at the event rainfall base. This choice is consistent with the objective of simulating runoff and erosion over time pans of more than 10 years as it ensures a good representativity of erosion at the catchment scale [49].

All the events generating effective runoff and corresponding to a duration above 3 h, volume above 40 mm and maximum 5-min intensity above 2 mm/h were selected and used in the calibration process. This selection includes both pre- and post-fire events generating together about 90% of the total measured soil erosion. Additionally, in order to inspect a priori the influence of the rainfall factors on runoff and erosion, a regression analysis was performed to relate runoff and SY with rainfall volume, antecedent rainfall, maximum intensity and duration of the event.

2.4. Climate and Fire-Related Scenarios

In this study, climate data were downloaded from the platform Medcordex.eu for both historical (1986–2005) and future (2041–2060) periods. Only the intermediate Representative Concentration Pathway (RCP) 4.5 climate emission scenario [50] was used because differences between RCP scenarios by 2050 were very small. The Regional Climate Model (RCM) ALADIN was chosen in the Mediterranean Agricultural Soils Conservation under global Change (MASCC) project; it was considered as a valid climate model in the comparison of RCM by Fantini et al. (2018) [51]. A statistical downscaling based on a quantile mapping approach was applied on the daily time rainfall series for bias correction, and a temporal downscaling based on an analog approach was then applied to transform the daily rainfall time series into five minutes' time series. Thus, the 5-min time series have a real temporal structure as a direct result of the observations.

Burnt area scenarios were calculated from climate scenarios according to Sousa et al. (2015) [21]. To this end, the annual burnt area in the region surrounding Macieira (the Águeda river basin) between 1990–2016 was calculated from Portuguese historical databases [13], and a multiple linear regression was calculated with monthly rainfall anomalies, extreme fire weather (days with daily severity rating (DSR) above the 90th percentile [52]) and fire history in the preceding years. Meteorological data was taken from the ECMWF Re-Analysis [53]. While the agreement with observations was not very high ($r^2 = 0.50$), it was sufficient to provide a "best guess" of how much and when the burned area would change for the climate scenarios. Burnt area scenarios under climate change were calculated using rainfall and DSR projections, the latter with a correction for bias using an empirical quantile mapping approach [54]. The estimated doubling of burnt area, reflecting an increase in fire frequency rather than the size of the burnt areas, was more conservative than the three- to four-fold increase obtained with the same methodology for northwestern Iberia [21]. The results for the Águeda were subsequently downscaled for Macieira watershed, by assuming a doubling of fire frequency, from 1× to 2× every 20 years. Fire location was that of the 2011 fire, while the burnt area was the same in all scenarios, except for S2 with 5% larger burn area due to the eucalypt stand in the southern part of the catchment (Figure 2).

2.5. Land Use and Management Scenarios

Four narrative storylines and land use scenarios were used in this study (Figures 3 and 4, Supplementary S1). For this, six local researchers working in different institutions of the Central Portugal NUTS II region (where Macieira is located) were interviewed and completed a survey on writing narrative scenario storylines for four contrasted socio-economics trajectories, initially, designated in the MASCC project: S1 (business-as-usual), S2 (market-oriented), S3 (environmental protection) and S4 (sustainable) in the context of global change in the Caramulo region (Supplementary S1, Tables S1–S3). The survey encompassed a table with future land use distribution, a table with allocation constraints per land use, and a ranking exercise on post-fire soil management rated by local experts. The main elements of the narratives of the four expert-based scenarios are shown in Figure 3.

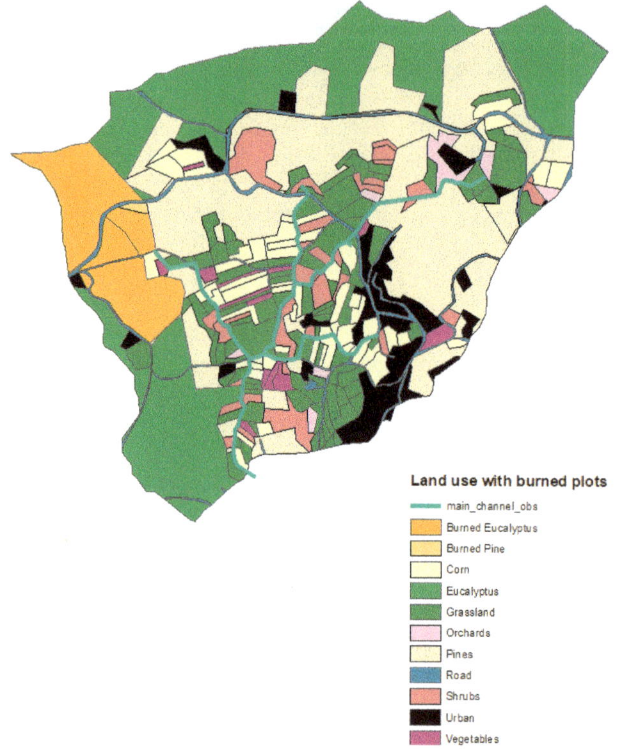

Figure 2. Location of the burned area in orange in our simulations corresponding to the area that burned in 2011.

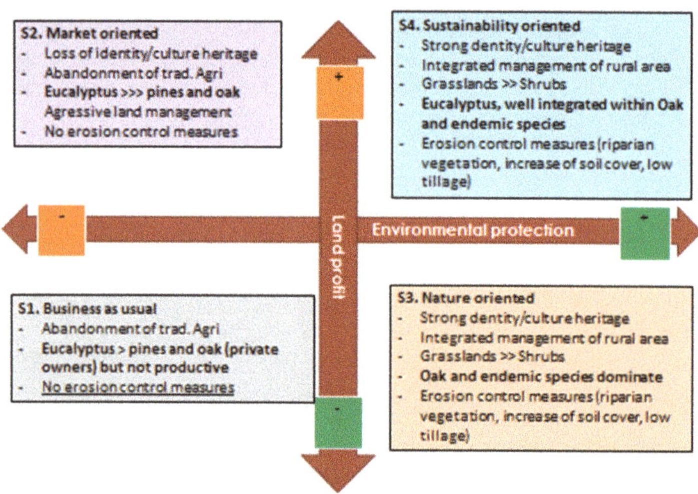

Figure 3. The four narrative land use scenarios elaborated under the MASCC project and used in this study.

Figure 4. Land use scenarios.

The current situation (S0) in Macieira is described as that of traditional agriculture (including corn and grassland) that has gradually been abandoned for afforestation, especially with *Eucalyptus* plantations that now occupy both abandoned agricultural areas and large areas of former pine forest and shrubland (Figure 4). Orchards, although still marginal, are increasing in all our scenarios as being a new profitable production system (berries). In all the scenarios, pine plantation decreases due to low profitability and due to reported increased pests and diseases, while oak is re-introduced in the environmental protection scenario (S3) and the sustainable scenario (S4) for biodiversity conservation and increase in soil water holding capacity, with incentives from local associations and/or government. In the S3 scenario, the maintenance and/or restoration of riparian vegetation is foreseen to capture sediments along the stream (with vegetated strips), in combination with adoption of conservation agriculture with lower tillage intensity. In S4, the same biodiversity and soil protection measures were implemented as in S3 but to only half the extent. In S3 and S4, soil tillage was also adjusted to follow contour lines.

As post-fire management, we selected application of mulch because it was suggested by local-experts as the most cost-effective post-fire conservation measure in this region, which is consistent with the findings of studies in the region [28,29,32,55] as well as in other parts of Iberian Peninsula [56,57].

2.6. Simulation Design

Model simulations concerned the period from 1986 to 2005 as well as a future period of 20 years between 2041 and 2060. They included the simulation of the exhaustive time series of major rainfall events, i.e., rainfall events with a mean lager than 40 mm. First, we assessed the impact of fire frequency and mulch application on runoff and soil erosion (Table 2) and then we assessed the combined impacts of land use and climate change scenarios, following the matrix in (Table 3).

Table 2. Impact of fire frequency and mulching on runoff and sediment yield (*SY*) on the baseline scenario S0.

No Climate Change (noCC)
No fire
1 fire
1 fire + Mulch
2 fires
2 fires + Mulch

Table 3. Impact of global change on runoff and sediment yield (*SY*).

	No Climate Change (noCC)	Climate Change (CC) Including 1 Fire	Climate Change (CC) Including 2 Fires
S0 Control S1 Business as usual S2 Market-oriented S3 Environmental protection S4 Sustainable	Period: 1986–2005 1 fire in 1995	Period: 2041–2060 1 fire in 2051	Period: 2041–2060 2 fires in 2042 and 2051

2.7. Data Analysis

The impact of fire frequency, mulch application and changes in climate and land use and management on runoff and erosion were analyzed by comparing the LANDSOIL simulations outputs with R studio software Version 1.0.136. In this paper, three levels of fire frequency were tested (i.e., 0, 1 and 2 fires every 20 years); two levels of mulch application (with or without application); two levels of climate change (with or without) and five scenarios of land use and management (S0 to S4). The simulation of the combinations of factors described in Tables 3 and 4 permitted to evaluate the distinct and combined impacts of these factors for the 2041–2060 period compared to the 1986–2005 period. A period of 3 post-fire years was also used to analyze the direct impact of fire and related contribution over 20 years. When there was no fire, we calculated the runoff and erosion produced during the period when fire would have had occurred to be able to compare with the period with fire. When two fires occurred, we calculated the runoff and *SY* that occurred two times: for three years after each fire.

Table 4. Impact of fire frequency and mulch application on runoff and sediment yield (*SY*) over 20 years (1986–2005). Control represents «S0 + 1 Fire» in 20 y, ii represents «no fire», iii represents «1 Fire including post-fire mulch application», iv represents «2 Fires» and v represents «2 Fires including post-fire mulch application». Relative values are in italics.

		Cumulative Runoff during 3 Post-Fire Years (mm)	Cumulative Runoff (mm)	Relative Contribution of Fire Events * to Runoff	SY during 3 Post-Fire Years (Mg ha^{-1})	SY (SY) (Mg ha^{-1})	Relative Contribution of Fire Events * to SY
Absolute values	Control (S0 + 1 fire) (i)	299	1948	15%	0.27	0.63	43%
	Control without fire (ii)	229	1878	12%	0.06	0.42	15%
	Control + mulch (iii)	279	1927	14%	0.17	0.53	32%
	2 fires (iv)	755	2042	37%	0.56	0.81	69%
	2 fires + mulch (v)	710	1992	36%	0.34	0.59	58%
Relative values (%)	Impact of preventing fire occurrence (ii − i)/i	−23%	−3%		−82%	−33%	
	Impact of doubling fire frequency (iv − i)/i	153%	5%		86%	28%	
	Impact of mulch if 1 fire (iii − i)/i	−7%	−1%		−35%	−16%	
	Impact of mulch if 2 fires (v − iv)/iv	−6%	−2%		−40%	−28%	

* Fire events represents the 3 years after the fire occurred.

3. Results

3.1. Variable Explaining Runoff and Sediment Yield in Macieira de Alcoba

Multiple linear regression analyses of event-wise runoff volumes (in m^3) and *SY*s (in Mg) gave satisfactory fits. This was especially true for runoff with an R^2 of 0.85 as opposed to 0.58 for *SY*. The best-fitting equations were:

$$Runoff = 581.12 \times Pt - 97.28 \times PI30 - 165.74 \times Pd + 213.34 \times AP48 - 20354.82, \tag{1}$$

$$SY = 49.46 \times Pt - 18.55 \times PI30 - 14.64 \times Pd + 131.97 \times AP48 - 2964.59, \qquad (2)$$

where: Pt is rainfall volume (in mm), $PI30$ is the maximum 30-min rainfall intensity (in mm.h^{-1}), Pd is the duration of the rainfall event and $AP48$ is the amount of rainfall over the 48 h preceding the event. The regressions results showed that rainfall and SY were explained best by Pt ($p < 0.001$), then by $AP48$ ($p < 0.01$) and less significantly Pd and $PI30$ (see Supplementary S1).

3.2. Calibration of Runoff and Sediment Yields Using the LandSoil Model (2010–2014)

The calibration results were obtained from creating two sets of infiltration rates according to the rainfall events (pre-fire and 3 years of post-fire disturbance) and by creating two sets of erosion by calibrating sediment concentration for diffuse erosion and rill dimension according to pre-fire conditions and fire conditions (three years following fire event) based on the observed data. The calibration of 30 events resulted in a satisfactory simulated runoff with R^2 of 0.92 and R^2 of 0.85 for simulated SY (Figure 5). We observed a better performance of the simulated runoff and sediment yield in the period post-fire than pre-fire however the period before fire had much lower values than after fire. The averages and SD were similar in both periods and variables. The NSE has the total value of 0.91 for runoff and 0.86 for sediment yield (see Supplementary S9).

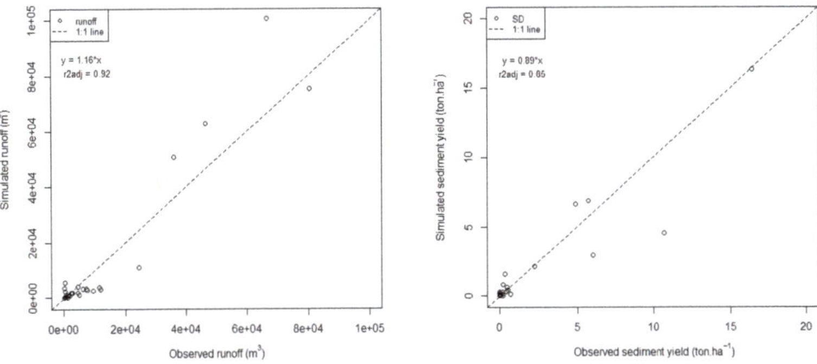

Figure 5. Results of calibration of LandSoil model for runoff (**left plot**) and sediment yield (**right plot**) with observed events between 2010 and 2013. SD stands for sediment yield.

3.3. Impact of Climate Change on Major Precipitation Events

Under the future climate conditions, the total number of major rainfall events over the 20-year period was 5% higher than under present climate conditions. Furthermore, the cumulative rainfall of these events was 7% higher under the future than present climate conditions (Figure 6 and Supplementary S1). At the same time, rainfall duration and maximum intensity, as well as antecedent rainfall, were between 15% and 30% higher under future climate conditions than under the current situation.

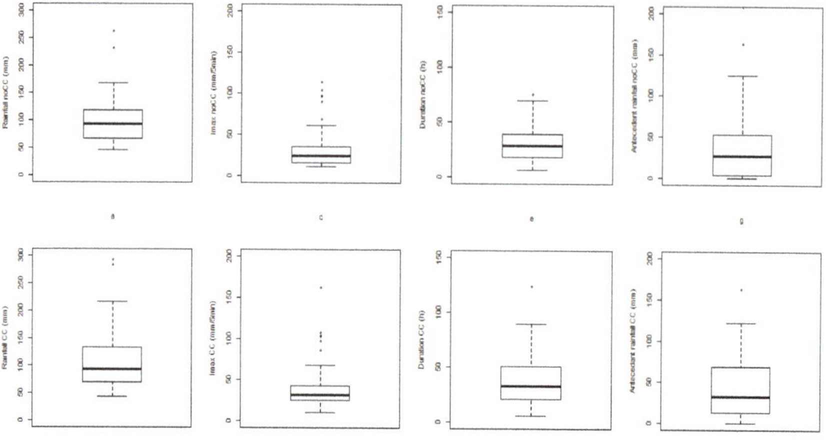

Figure 6. Boxplots of key characteristics of major rainfall events under present (no climate change—noCC, 1986–2005) and future (climate change—CC, 2041–2060) climate conditions: rainfall total (**a,b**), maximum intensity over 5 min (**c,d**), duration (**e,f**) and antecedent 48 h rainfall (**g,h**).

3.4. Impacts of Fire Frequency and Post-Fire Mulching on Erosion under Present Climate Conditions

Compared to the control scenario (one fire over the 20 years period 1986–2005), the scenario without fire reduced runoff by 3% over 20 years and by 23% during the 3 years of post-fire disturbance while the scenario with two fires increased runoff by 5% over 20 years and by 153% during the 2 × 3 years of post-fire disturbance (Table 4). In turn, post-fire mulching reduced runoff by 1% and 2% in the case of one and two fires over 20 years, respectively and up to 7% during the post-fire period only. Sediment yield was more strongly affected by fire frequency and post-fire erosion mitigation than runoff. The scenario without fire decreased SY by 33% over the 20 year period (and by 82% during the post-fire period only) compared to the control scenario, while the one with two fires increased it by 28% over the 20 years and by 86% (during the post-fire period only). The scenarios involving post-fire mulch application decreased SY by 16% and 28% in the case of one and two fires, respectively (and by 35% and 40% during the post-fire period only). While with one fire, the years of fire disturbance represents 15% of the runoff generation, when doubling fire frequency, the two periods of fire disturbance generation about 37% of runoff. Concerning SY, under the control situation, the fire disturbance period represents 43% of the SY but when fire frequency doubles, 69% of the sediments are produced during both fire periods (six years out of 20). In the scenario without fire, we showed that during the 3 years when fire would have occurred; only 12% and 15% of runoff and SY were respectively produced. Finally, the contribution of interrill erosion to total erosion accounted for 1%–8% of the total erosion in all the simulations.

3.5. Runoff and Sediment Yield Under Future Climate Conditions

Without changes in fire frequency, cumulative runoff under future climate conditions was 9%–21% higher than under present climate conditions (Table 5, Figures 7 and 8). With a doubling of fire frequency, the cumulative runoff was 13%–25% higher under the future than present climate conditions. The impacts of changes in climate conditions, as well as fire frequency, were most pronounced in the case of the business-as-usual (S1) and market-oriented scenarios (S2). Land use had reduced impacts on runoff under present climate conditions (with differences of up to 3%, in the case of S2) but marked impacts under future climate conditions, with runoff differences amounting to 11%–13% for the S3 and S4 scenarios and to 22%–23% for the S1 and S2 scenarios.

Table 5. Impact of fire frequency under land use and management and climate change on cumulative runoff and sediment yield (SY) over 20 years. Runoff is in mm and sediment yield in Mg ha^{-1}. The impact is quantified as a percentage of change relative to the control. noCC represents no climate change within the 1986–2005 period and CC represents climate change within the 2041–2060 period.

Parameter		Runoff (mm)					Sediment Yield (Mg ha^{-1})				
Scenario		S0	S1	S2	S3	S4	S0	S1	S2	S3	S4
Absolute values	Control (noCC + 1 fire) (i)	2046	2064	2098	2029	2044	0.61	0.80	1.17	0.29	0.47
	CC including 1 fire (ii)	2227	2489	2513	2262	2311	0.73	1.17	1.65	0.35	0.52
	CC including 2 fires (iii)	2314	2576	2612	2340	2465	0.90	1.38	1.78	0.46	0.65
Relative changes	Impact of CC if 1 fire ((ii) − (i))/(i)	9%	21%	20%	11%	13%	19%	45%	41%	21%	11%
	Impact of CC if 2 fires ((iii) − (i))/(i)	13%	25%	24%	15%	21%	48%	73%	52%	59%	38%
	Impact of fire under CC ((iii) − (ii))/(ii)	4%	3%	4%	3%	7%	23%	18%	8%	29%	12%
	Impact of LU under present climate conditions (comparison of Sx − S0 /S0 for (i))		1%	3%	−1%	0%		31%	92%	−52%	−23%
	Impact of LU under CC (comparison of Sx − S0 /S0 for (iii))		11%	13%	1%	7%		53%	98%	−49%	−28%
	Combined impact of LU + CC (with 2 fires) (comparison of S1–S4 including CC (iii) to S0 without CC)	13%	26%	28%	14%	20%	48%	126%	192%	−25%	7%

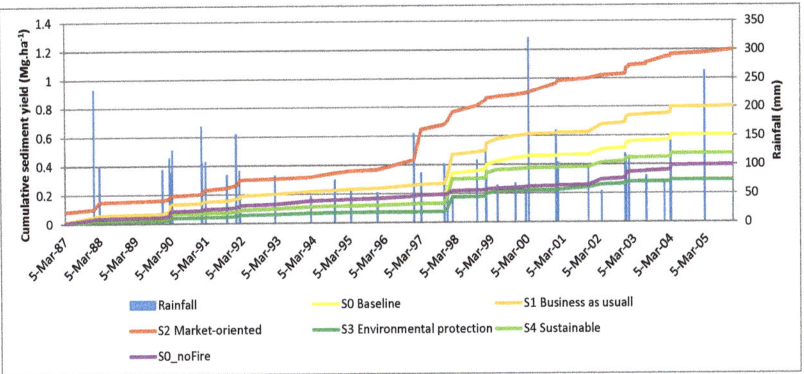

Figure 7. Cumulative sediment yield (SY) in Mg ha^{-1} vs. rainfall in mm between 1986 and 2005 with one fire occurrence in 1995.

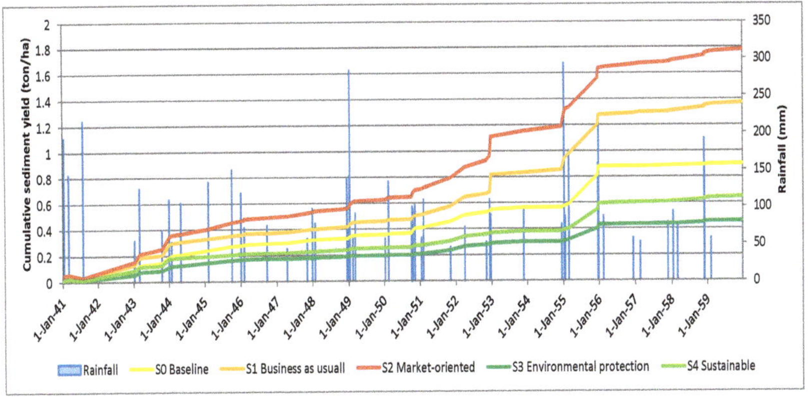

Figure 8. Cumulative sediment yield (SY) in Mg ha^{-1} vs. rainfall in mm under climate change between 2041 and 2060 with two fire occurrences in 1941 and 1951.

Concerning sediment yield (Table 5), we observed that the impact of climate change increased *SY* from 11% (S4 scenario) up to 45% in the S1 scenario under one fire and the impact of increasing fire frequency could increase *SY* up to 29%. Concerning the impact of land use, we observed high contrasts between scenarios. Without climate change, *SY* increased from 32% in S1 up to 90% in S2 while we observed a decrease in *SY* in S3 and S4, respectively by 52% and 21%. The combined impacts of climate change and land use change show an increase of *SY* by 60% and 126% in S1 and S2 respectively and a decrease by 22% and 47% in S4 and S3 respectively. Land use change and management have a strong impact on *SY* with nearly a doubling of *SY* in S2 and a reduction by half in S3 scenario. Figure 9 shows the net cumulative erosion (soil loss) and deposition (soil gain) between 1986 and 2005 for the S2 scenario. Three fields, including two burned fields, show the highest cumulative soil loss between 7 and 9 Mg ha^{-1}, while the areas in the north of the catchment show net deposition probably coming from the road located on the NW of the catchment. In Figure 10, the impact of increasing riparian vegetation in S3 shows the impact of sediment deposition around the stream. Overall, we show that the highest negative impact would be to change the landscape to mainly *Eucalyptus* plantations with no conservative land management approaches (S2); while with sustainable management (S3) *SY* can be reduced by half.

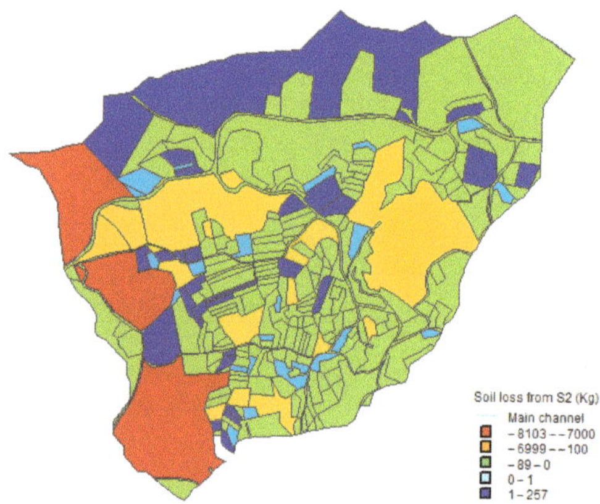

Figure 9. Soil loss between 1986 and 2005 in S2 after two fires (Kg). Negative values represent soil loss and positive values represent gain.

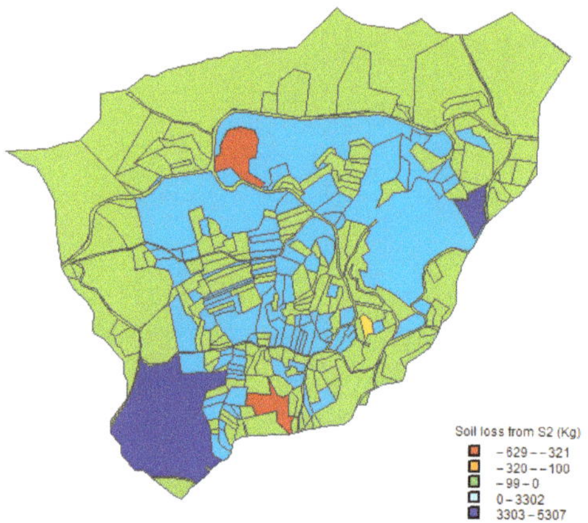

Figure 10. Soil amount difference between Scenario S3 and S0 without a fire at year 2005 (Kg). Negative values represent loss and positive values represent gain.

4. Discussion

4.1. Application of LandSoil to Macieira

In our study, we found that the volume of rainfall during and 48 h before the fire event was the most important climatic driver of the catchment-scale soil erosion. From these, rainfall volume had the strongest correlation with erosion rates and this can be due to the humid Mediterranean climate in Macieira de Alcoba, with usually bigger and longer events than in drier Mediterranean regions, and concurs with the results of similar other studies in NW Iberia [58]. The calibration of the infiltration rate, sediment concentration and rill dimension led to a good performance of LandSoil in representing current runoff and *SY*. The only land use that had to be calibrated according to literature was the re-introduction of oak forest, which leads in time to a higher soil water infiltration than Eucalypt [59]). Performance of LandSoil might be improved with an extended field measurement dataset but this would have required more field monitoring and probably another fire occurrence with a different extent and location in the catchment.

4.2. Overall LandSoil Results

The model application to Macieira indicates that the impact of climate change is relatively low, increasing runoff generation by 9% and sediment yield by 19%. Doubling fire frequency has low impacts on runoff in the long-term, increasing by 5%, but larger impacts on *SY*, which increases by 28%. However, the model also shows the effectiveness of mulch application after fires on *SY* with a reduction of 16%–28% with one and two fires respectively (and up to 40% reduction during the post-fire disturbance period only). The application of mulch had a lesser impact on runoff than on *SY* (up 2% reduction) over 20 years and up to 7% during the post-fire disturbance period. Finally, the model indicates that land use change had the lowest impact on runoff, with changes between −1% and 3%, but the largest impacts on *SY*, changing between −23% and 92%. These results show that land use allocation and management was the most important factor impacting *SY* with a potential exacerbation of fire and climate change impacts in the case of S1 and S2 scenarios and strong potential of mitigation in S3 and S4 scenarios. In these latter scenarios, a large part of sediments was trapped in riparian fields, therefore preventing sediments to reach the stream and outlet (Figure 10).

This decreased *SY* more than soil erosion; while this would not mitigate land degradation in burned areas, it would limit stream contamination by ash transport after a fire event, which can have significant impacts on water resources [19]. Compared to the modeling study by Nunes et al. (2018b), our results showed less impact of fire on runoff and *SY*; while in Nunes et al. (2018a) [60] a 'no fire situation' decreases *SY* by 2/3, in our case *SY* was decreased by 1/3. This could be due to the duration of the calculated cumulative *SY*. In Nunes et al. (2018a, b), this is done for a 10-yr period as opposed to for a 20 years period in this study. In fact, when our results were only compared during the three years of post-fire disturbance, the efficiency of mulch application decreased runoff by 7% and *SY* by 40%. Compared to Prats et al. (2019, 2012) where mulch application reduced runoff by 15%–25% and *SY* by more than five times, our results were more conservative probably because our burnt field area only represented 10% of the watershed area; while Prats et al. (2012, 2019) calculated *SY* at the field area scale and they burned 100% of the area while in our case 10% of the catchment burned. Moreover, our simulations ran over 20 years vs. 5–10 years in other studies [32,61].

Our study shows the low impact of mulch application on runoff generation. This could be due to high soil water repellency in the eucalypt burnt forest. Soil water repellency also exists in *Eucalyptus* forests and in a pre-fire situation due to preponderant ligneous root materials compared to oak and pine trees [59,62]. However, soil water repellency has a limited impact on runoff generation, in the early autumn [61]; and this is only important for patches since at the hillslope scale there are always non-repellent zones where water can re-infiltrate [15]. The low impact of mulching on runoff reduction is also probably due to the low total amount of runoff generation in both burnt and unburnt eucalypt plantations, at least at the hillslope scale [15,37]. Finally, the risk of increasing floods remains relatively low compared to predicted extreme rainfall events [60] due to the low impact of fires on runoff production.

Concerning sediment reduction caused by mulching, Shakesby et al. (1996) [63], showed that *Eucalyptus* logging litter reduced soil erosion by 95%. The average reduction was five times in Prats et al. (2019, 2012); it should be noted that, in the latter study, maximum erosion rates were lower than in our study site. As previously said, the study site of Prats et al. (2012) also presents smaller temporal and spatial scales. In our study, mulch could reduce SY by up to 40%, which was quite high considering that fire only impacted roughly 10% of the watershed. We also suggest that these values vary with the type and quantity of mulch that is applied, the soil type, the percentage of soil cover and the slope of the site (Prats et al., 2014, 2012). Finally, Figures 5 and 6 show that erosion was mainly a raindrop-driven process with the detachment of sediments in eucalypt forest and some tilled field. The highest levels of soil loss occurred in the burnt field area and in one field with a slope above 10° in the S2 scenario (about 7–8 Mg per parcel) and these three areas contribute to 37% of total soil loss in scenario S2 (Figures 7 and 8). The combined impact of fire on SY at the outlet and the spatial representation of soil loss coming from burned areas show that erosion was mainly due to soil detachment rather than transported by water to the outlet via sedimentation on the riparian field close to the stream and in the terraces.

4.3. Distinct and Combined Impacts of Global Change and Fire Frequency

In light of future climate change, it is likely that fire frequency could increase in a context of climate change and that doubling fire frequency might be conservative in our study when compared with burned area projections for this region [21]. The land use scenarios might also impact fire frequency and extent; for example, the increase of *Eucalyptus* plantation and intensification in S1 and S2 might exacerbate the risk of fire occurrence and propagation, while in S3 and S4 the adoption of crop diversification and protective land use and management should lead to lower risks of fires and erosion. However, we did not take these differences into account because we wanted to have the same number of fires in our land use scenarios so the specific impact of fire would be comparable. In any case, our results show that fire frequency had a larger impact on soil erosion (up to 33% change) than on runoff generation (up to 5% change). In addition, our results also show 43% of the erosion was produced during 3 years after the fire event representing 15% of the duration of the simulation (Table 4). This concurred with multiple studies, which showed that the indirect impacts of climate change, such as associated changes to land use and fire occurrence can have larger impacts than those of changes to climatic variables alone [9,64–66]. In the fire-prone forests of NW Iberia, efforts to limit the impact of climate change on land degradation should, therefore, focus on fire prevention and post-fire impact mitigation.

4.4. Status of Post Fire Measures Implementation

Our study shows the potential of using soil management measures to mitigate post-fire erosion in NW Portugal with a mix of numerical modeling and field experiments. The biophysical effectiveness of erosion control measures is well understood in NW Portugal [67,68]. Studies show the high performance of mulching and erosion control barriers, and the lesser effectiveness of erosion control measures such as post-fire seedling, which are more difficult to manage due to the time lag of seed germination and the availability of native seeds [69]. However, there is currently only one study calculating the costs of post-fire control erosion measures [68]. The benefits produced by controlling overland runoff and SY with control erosion measures on hydrological services should be biophysically and economically quantified so that implementation of erosion control measures can be communicated at the science-policy management level [70]. For instance, the risk of water contamination must be studied and costs of prevention (with erosion control measures) compared to the costs of water treatment must be evaluated [18,60,71,72]. The same applies to the hydrological service of flood regulation and enhanced water soil content and fertility with post-fire erosion control measures via increased soil organic matter, soil aggregate stability and nutrient availability [73]. For this work to proceed, more studies are needed on the socio-economic aspects of fire prevention and post-fire contamination control.

5. Conclusions

This study evaluated the potential impacts of global change and fire frequency on water runoff and soil erosion. The novelty of this study lies with the assessment of global change with field and landscape conservation management techniques such as application of mulch in post-fire conditions, implementation of vegetated strips (or restoration of riparian vegetation) and decreased soil tillage on water runoff and soil erosion. Our results show that the LandSoil model performed well for runoff and *SY* compared to the observed data of 2010–2014 and that rainfall volumes (>40 mm) during and 48 h before fire occurrence are the main climate drivers of soil erosion before rainfall duration and maximum intensity. With LandSoil we could test contrasted land use scenarios including *Eucalyptus* intensification to show their potential risk of implementation under global change (up to +98% in *SY* in the market-oriented (S2) scenario) while a conservation scenario including environmental protection and afforestation of local species can decrease soil erosion by half. The application of mulching was an effective tool against erosion (up to 28% with two fires in 20 years). Finally, preventing fire is also a potential erosion mitigation tool with a decrease of 33% of soil erosion. Valuating hydrological services of post-fire erosion measures would clarify the feasibility of implementation of these latter by communicating the results within a science-policy interface in a context of the latest Intergovernmental Panel on Climate Change (IPCC) recommendations on achieving climate change mitigation and land degradation neutrality (LDN).

Supplementary Materials: The following are available online at http://www.mdpi.com/2073-4441/11/12/2617/s1, Supplementary S1: Narrative Storyline of Land Use Scenario of MASCC Project, Supplementary S2: Land Use and Climate Scenarios. Table S1: Land use allocation constrains for Macieira, Table S2: Land use allocation per scenario S0 (baseline 2000), S1, S2, S3 and S4, Table S3. Soil conservation practices ranking by six local experts during MASCC project, Table S4: Statistical summary of rainfall events selected on the study site during the period 1986–2005 (noCC) and 2041–2060 (CC), Supplementary S3: Results from Statistical Analyses of Observed Rainfall Events and Corresponding Runoff and Sediment Yield (R version 3.4.2 (28 September 2017)—"Short Summary"), Supplementary S4: Table S4: Calibration of Monthly Calendars of Soil Roughness, Soil Crusting and Soil Cover per Land Use, Supplementary S5: Soil Tillage Calibration, Supplementary S6: LandSoil Turbidity Calibration, Supplementary S7: Calibration of Rill Erosion, Supplementary S8: Calibration of Infiltration Rates for Two Periods (Pre-Fire and during the Fire Disturbance Period), Supplementary S9: Results from the Calibration of Runoff and Sediment Yield in Pre and Post Fire Periods. Figure S1: Difference in elevation change between scenario S3 and S0 without fire occurrence. Figure S2: Elevation change in scenario S2 between 2000 and 2050.

Author Contributions: Conceptualization: A.V.P., J.P.N., R.C., Y.L.-B., J.J.K. and D.R.; Formal analysis: A.V.P., J.P.N., M.K. and F.H.; Funding acquisition: T.C. and D.R.; Investigation: J.P.N., Y.L.-B., J.J.K. and D.R.; Methodology, A.V.P., J.B., F.H., T.C., Y.L.-B. and D.R.; Software: R.C. and M.K.; Supervision: J.P.N., J.B. and Y.L.-B.; Validation: J.P.N. and D.R.; Visualization: A.V.P.; Writing—original draft: A.V.P., J.P.N., D.R.; Writing—review & editing, J.P.N., R.C., M.K., J.B., F.H., T.C., J.J.K. and D.R.

Funding: This work benefits from the financial support of the MASCC project through ARIMNET2, an ERA-NET funded by the European Union's Seventh Framework Program for research, technological development and demonstration under grant agreement no. 618127. APV was funded by the MASCC project described above. This work was further funded by the "Fundação para a Ciência e a Tecnologia", with personal grants attributed to J.P. Nunes (IF/00586/2015) and J.J. Keizer (IF/01465/2015), and funding for data collection in Macieira within project ERLAND (FCOMP-01-0124-FEDER-008534). J. Baartman visited the site with funding from an STSM mission within the Connecteur COST Action ES1306. M. Koopmans was partly funded by ERASMUS+.

Acknowledgments: We would like to thank the two external reviewers and the editorial board for their valuable comments before the acceptance of the manuscript.

Conflicts of Interest: The authors declare no conflict of interest.

References

1. Lal, R. Accelerated soil erosion as a source of atmospheric CO_2. *Soil Tillage Res.* **2019**, *188*, 35–40. [CrossRef]

2. Walker, B. Global change extensive strategy agriculture options in the regions of the world. *Clim. Chang.* **1994**, *27*, 39–47. [CrossRef]

3. Montgomery, D.R. Soil erosion and agricultural sustainability. *Proc. Natl. Acad. Sci. USA* **2007**, *104*, 13268–13272. [CrossRef] [PubMed]

4. Vanmaercke, M.; Poesen, J.; Verstraeten, G.; de Vente, J.; Ocakoglu, F. Sediment yield in Europe: spatial patterns and scale dependency. *Geomorphology* **2011**, *130*, 142–161. [CrossRef]

5. Woodward, J.C. *Patterns of Erosion and Suspended Sediment Yield in Mediterranean River Basins*; Wiley: Chichester, UK, 1995.

6. Cerdan, O.; Govers, G.; Le Bissonnais, Y.; Van Oost, K.; Poesen, J.; Saby, N.; Gobin, A.; Vacca, A.; Quinton, J.; Auerswald, K. Rates and spatial variations of soil erosion in Europe: a study based on erosion plot data. *Geomorphology* **2010**, *122*, 167–177. [CrossRef]

7. Panagos, P.; Borrelli, P.; Meusburger, K.; Alewell, C.; Lugato, E.; Montanarella, L. Estimating the soil erosion cover-management factor at the European scale. *Land Use Policy* **2015**, *48*, 38–50. [CrossRef]

8. De Franchis, L.; Bleu, P.; Ibanez, F. *Threats to Soils in Mediterranean Countries: Document Review*; PAM, Plan Bleu: Sophia-Antipolis, France, 2003.

9. Raclot, D.; Le Bissonnais, Y.; Annabi, M.; Sabir, M.; Smetanova, A. Main issues for preserving Mediterranean soil resources from water erosion under global change. *L. Degrad. Dev.* **2018**, *29*, 789–799. [CrossRef]

10. García-Ruiz, J.M.; Nadal-Romero, E.; Lana-Renault, N.; Beguería, S. Erosion in Mediterranean landscapes: changes and future challenges. *Geomorphology* **2013**, *198*, 20–36. [CrossRef]

11. Shakesby, R.A. Post-wildfire soil erosion in the Mediterranean: review and future research directions. *Earth Sci. Rev.* **2011**, *105*, 71–100. [CrossRef]

12. Batllori, E.; Parisien, M.; Krawchuk, M.A.; Moritz, M.A. Climate change-induced shifts in fire for Mediterranean ecosystems. *Glob. Ecol. Biogeogr.* **2013**, *22*, 1118–1129. [CrossRef]

13. Pereira, M.G.; Malamud, B.D.; Trigo, R.M.; Alves, P.I. The history and characteristics of the 1980–2005 Portuguese rural fire database. *Nat. Hazards Earth Syst. Sci.* **2011**, *11*, 3343–3358. [CrossRef]

14. Doerr, S.H.; Santín, C. Global trends in wildfire and its impacts: perceptions versus realities in a changing world. *Philos. Trans. R. Soc. B Biol. Sci.* **2016**, *371*, 20150345. [CrossRef] [PubMed]

15. Ferreira, A.J.D.; Coelho, C.O.A.; Ritsema, C.J.; Boulet, A.K.; Keizer, J.J. Soil and water degradation processes in burned areas: lessons learned from a nested approach. *Catena* **2008**, *74*, 273–285. [CrossRef]

16. Ferreira, A.J.D.; Alegre, S.P.; Coelho, C.O.A.; Shakesby, R.A.; Páscoa, F.M.; Ferreira, C.S.S.; Keizer, J.J.; Ritsema, C. Strategies to prevent forest fires and techniques to reverse degradation processes in burned areas. *Catena* **2015**, *128*, 224–237. [CrossRef]

17. Campos, I.; Abrantes, N.; Pereira, P.; Keizer, J.J. *Chap. 9. Polycyclic aromatic hydrocarbons*; Pereira, P., Mataix-Solera, J., Úbeda, X., Rein, G., Cerda, A., Eds.; CSIRO PUBLISHING: Sydney, Australia, 2019.

18. Nunes, B.; Silva, V.; Campos, I.; Pereira, J.L.; Pereira, P.; Keizer, J.J.; Gonçalves, F.; Abrantes, N. Off-site impacts of wildfires on aquatic systems—biomarker responses of the mosquitofish Gambusia holbrooki. *Sci. Total Environ.* **2017**, *581*, 305–313. [CrossRef]

19. Nunes, J.P.; Naranjo Quintanilla, P.; Santos, J.M.; Serpa, D.; Carvalho-Santos, C.; Rocha, J.; Keizer, J.J.; Keesstra, S.D. Afforestation, subsequent forest fires and provision of hydrological services: A model-based analysis for a Mediterranean mountainous catchment. *Land Degrad. Dev.* **2018**, *29*, 776–788. [CrossRef]

20. Bedia, J.; Herrera, S.; Gutiérrez, J.M.; Benali, A.; Brands, S.; Mota, B.; Moreno, J.M. Global patterns in the sensitivity of burned area to fire-weather: Implications for climate change. *Agric. For. Meteorol.* **2015**, *214*, 369–379. [CrossRef]

21. Sousa, P.M.; Trigo, R.M.; Pereira, M.G.; Bedia, J.; Gutiérrez, J.M. Different approaches to model future burnt area in the Iberian Peninsula. *Agric. For. Meteorol.* **2015**, *202*, 11–25. [CrossRef]

22. Barbero, R.; Abatzoglou, J.T.; Larkin, N.K.; Kolden, C.A.; Stocks, B. Climate change presents increased potential for very large fires in the contiguous United States. *Int. J. Wildl. Fire* **2015**, *24*, 892–899. [CrossRef]

23. Wang, X.; Thompson, D.K.; Marshall, G.A.; Tymstra, C.; Carr, R.; Flannigan, M.D. Increasing frequency of extreme fire weather in Canada with climate change. *Clim. Chang.* **2015**, *130*, 573–586. [CrossRef]

24. Liu, Z.; Wimberly, M.C. Direct and indirect effects of climate change on projected future fire regimes in the western United States. *Sci. Total Environ.* **2016**, *542*, 65–75. [CrossRef] [PubMed]

25. Syphard, A.D.; Sheehan, T.; Rustigian-Romsos, H.; Ferschweiler, K. Mapping future fire probability under climate change: Does vegetation matter? *PLoS ONE* **2018**, *13*, e0201680. [CrossRef] [PubMed]

26. Fernández, C.; Vega, J.A. Evaluation of RUSLE and PESERA models for predicting soil erosion losses in the first year after wildfire in NW Spain. *Geoderma* **2016**, *273*, 64–72. [CrossRef]

27. Ferreira, V.; Panagopoulos, T. Seasonality of Soil Erosion Under Mediterranean Conditions at the Alqueva Dam Watershed. *Environ. Manag.* **2014**, *54*, 67–83. [CrossRef] [PubMed]

28. Keizer, J.J.; Silva, F.C.; Vieira, D.C.S.; González-Pelayo, O.; Campos, I.; Vieira, A.M.D.; Valente, S.; Prats, S.A. The effectiveness of two contrasting mulch application rates to reduce post-fire erosion in a Portuguese eucalypt plantation. *Catena* **2018**, *169*, 21–30. [CrossRef]

29. Prats, S.A.; Malvar, M.C.; Vieira, D.C.S.; MacDonald, L.; Keizer, J.J. Effectiveness of Hydromulching to Reduce Runoff and Erosion in a Recently Burnt Pine Plantation in Central Portugal. *Land Degrad. Dev.* **2016**, *27*, 1319–1333. [CrossRef]

30. Gómez-Rey, M.X.; Couto-Vázquez, A.; García-Marco, S.; Vega, J.A.; González-Prieto, S.J. Reduction of nutrient losses with eroded sediments by post-fire soil stabilisation techniques. *Int. J. Wildl. Fire* **2013**, *22*, 696–706. [CrossRef]

31. Prats, S.A.; González-Pelayo, Ó.; Silva, F.C.; Bokhorst, K.J.; Baartman, J.E.M.; Keizer, J.J. Post-fire soil erosion mitigation at the scale of swales using forest logging residues at a reduced application rate. *Earth Surf. Process. Landf.* **2019**. [CrossRef]

32. Prats, S.A.; MacDonald, L.H.; Monteiro, M.; Ferreira, A.J.D.; Coelho, C.O.A.; Keizer, J.J. Effectiveness of forest residue mulching in reducing post-fire runoff and erosion in a pine and a eucalypt plantation in north-central Portugal. *Geoderma* **2012**, *191*, 115–124. [CrossRef]

33. Robichaud, P.R.; Ashmun, L.E.; Sims, B.D. *Post-Fire Treatment Effectiveness for Hillslope Stabilization*; General Technical Report RMRS-GTR-240; US Department of Agriculture, Forest Service, Rocky Mountain Research Station: Fort Collins, CO, USA, 2010.

34. Vega, J.A. *Acciones urgentes contra la erosión en áreas forestales quemadas: Guía para su planificación en Galicia*; INIA. Xunt.; Andavira: Santiago de Compostela, Spain, 2013; ISBN 8484087166.

35. Hosseini, M.; Keizer, J.J.; Pelayo, O.G.; Prats, S.A.; Ritsema, C.; Geissen, V. Effect of fire frequency on runoff, soil erosion, and loss of organic matter at the micro-plot scale in north-central Portugal. *Geoderma* **2016**, *269*, 126–137. [CrossRef]

36. Ferreira, C.G. Erosão Hídrica em Solos Florestais. Estudo em povoamentos de Pinus pinaster e Eucalyptus globulus em Macieira de Alcoba. 1996. Available online: http://www.dcs.ufla.br/site/_adm/upload/file/pdf/Prof%20Marx/Aula%205/Outro%20art%20interesse/Ferreira%201996_erosao%20em%20floresta.pdf (accessed on 10 December 2019).

37. Hawtree, D.; Nunes, J.P.; Keizer, J.J.; Jacinto, R.; Santos, J.; Rial-Rivas, M.E.; Boulet, A.-K.; Tavares-Wahren, F.; Feger, K.-H. Time series analysis of the long-term hydrologic impacts of afforestation in the Águeda watershed of north-central Portugal. *Hydrol. Earth Syst. Sci.* **2015**, *19*, 3033–3045. [CrossRef]

38. Nunes, J.P.; Bernard-Jannin, L.; Rodríguez Blanco, M.L.; Santos, J.M.; Coelho, C.O.A.; Keizer, J.J. Hydrological and erosion processes in terraced fields: observations from a humid Mediterranean region in Northern Portugal. *Land Degrad. Dev.* **2018**, *29*, 596–606. [CrossRef]

39. Ciampalini, R.; Follain, S.; Cheviron, B.; Le Bissonnais, Y.; Couturier, A.; Moussa, R.; Walter, C. Local Sensitivity Analysis of the LandSoil Erosion Model Applied to a Virtual Catchment. In *Sensitivity Analysis in Earth Observation Modelling*; Elsevier: Wales, UK, 2017; pp. 55–73.

40. Ciampalini, R.; Follain, S.; Le Bissonnais, Y. LandSoil: a model for analysing the impact of erosion on agricultural landscape evolution. *Geomorphology* **2012**, *175*, 25–37. [CrossRef]

41. Jenson, S.K.; Domingue, J.O. Extracting topographic structure from digital elevation data for geographic information system analysis. *Photogramm. Eng. Remote Sens.* **1988**, *54*, 1593–1600.

42. Souchere, V.; King, D.; Daroussin, J.; Papy, F.; Capillon, A. Effects of tillage on runoff directions: consequences on runoff contributing area within agricultural catchments. *J. Hydrol.* **1998**, *206*, 256–267. [CrossRef]

43. Souchere, V.; Cerdan, O.; Ludwig, B.; Le Bissonnais, Y.; Couturier, A.; Papy, F. Modelling ephemeral gully erosion in small cultivated catchments. *Catena* **2003**, *50*, 489–505. [CrossRef]

44. Cerdan, O.; Le Bissonnais, Y.; Couturier, A.; Saby, N. Modelling interrill erosion in small cultivated catchments. *Hydrol. Process.* **2002**, *16*, 3215–3226. [CrossRef]

45. Cerdan, O.; Souchère, V.; Lecomte, V.; Couturier, A.; Le Bissonnais, Y. Incorporating soil surface crusting processes in an expert-based runoff model: sealing and transfer by runoff and erosion related to agricultural management. *Catena* **2002**, *46*, 189–205. [CrossRef]

46. Govers, G.; Vandaele, K.; Desmet, P.; Poesen, J.; Bunte, K. The role of tillage in soil redistribution on hillslopes. *Eur. J. Soil Sci.* **1994**, *45*, 469–478. [CrossRef]

47. Van Muysen, W.; Govers, G.; Van Oost, K.; Van Rompaey, A. The effect of tillage depth, tillage speed, and soil condition on chisel tillage erosivity. *J. Soil Water Conserv.* **2000**, *55*, 355–364.

48. Hutchinson, M.F.; Xu, T.; Stein, J.A. Recent progress in the ANUDEM elevation gridding procedure. *Geomorphometry* **2011**, *2011*, 19–22.

49. Nunes, J.P.; Nearing, M.A. Modelling impacts of climatic change: case studies using the new generation of erosion models. Handbook of Erosion Modelling; John Wiley & Sons, Ltd.: Hoboken, NJ, USA, 2011; pp. 289–312.

50. Van Vuuren, D.P.; Edmonds, J.; Kainuma, M.; Riahi, K.; Thomson, A.; Hibbard, K.; Hurtt, G.C.; Kram, T.; Krey, V.; Lamarque, J.-F. The representative concentration pathways: an overview. *Clim. Chang.* **2011**, *109*, 5–31. [CrossRef]

51. Fantini, A.; Raffaele, F.; Torma, C.; Bacer, S.; Coppola, E.; Giorgi, F.; Ahrens, B.; Dubois, C.; Sanchez, E.; Verdecchia, M. Assessment of multiple daily precipitation statistics in ERA-Interim driven Med-CORDEX and EURO-CORDEX experiments against high resolution observations. *Clim. Dyn.* **2018**, *51*, 877–900. [CrossRef]

52. Van Wagner, C.E. *Development and Structure of the Canadian Forest Fire Weather Index System*; Canadian Forestry Service: Ottawa, ON, Canada, 1987; Volume 35, ISBN 0662151984.

53. Berrisford, P.; Dee, D.P.; Poli, P.; Brugge, R.; Fielding, M.; Fuentes, M.; Kallberg, P.W.; Kobayashi, S.; Uppala, S.; Simmons, A. *The ERA-Interim archive Version 2.0*; ECMWF: Reading, UK, 2011.

54. Gudmundsson, L.; Bremnes, J.B.; Haugen, J.E.; Engen-Skaugen, T. Downscaling RCM precipitation to the station scale using statistical transformations–a comparison of methods. *Hydrol. Earth Syst. Sci.* **2012**, *16*, 3383–3390. [CrossRef]

55. Prats, S.; Malvar, M.; Martins, M.A.S.; Keizer, J.J. Post-fire soil erosion mitigation: a review of the last research and techniques developed in Portugal. *Cuad. Investig. Geográfica* **2014**, *40*, 403–428. [CrossRef]

56. Badia, D.; Marti, C. Seeding and mulching treatments as conservation measures of two burned soils in the central ebro valley, ne spain. *Arid Soil Res. Rehabil.* **2000**, *14*, 219–232. [CrossRef]

57. Fernández, C.; Vega, J.A. Are erosion barriers and straw mulching effective for controlling soil erosion after a high severity wildfire in NW Spain? *Ecol. Eng.* **2016**, *87*, 132–138. [CrossRef]

58. Rodriguez-Blanco, M.L.; Taboada-Castro, M.M.; Palleiro, L.; Taboada-Castro, M.T. Assessment of the effects of rainfall variability on the hydrological regime of a small rural catchment in Northwest Spain: preliminary results. In Proceedings of the EGU General Assembly Conference Abstracts, Vienna, Austria, 2–7 May 2010; Volume 12, p. 14227.

59. Alves, A.M. *O eucaliptal em Portugal: impactes ambientais e investigação científica*; Univ. Técnica, Inst. Superior de Agronomia: Lisbon, Portugal, 2007; ISBN 9728669259.

60. Nunes, J.P.; Doerr, S.H.; Sheridan, G.; Neris, J.; Santín, C.; Emelko, M.B.; Silins, U.; Robichaud, P.R.; Elliot, W.J.; Keizer, J. Assessing water contamination risk from vegetation fires: challenges, opportunities and a framework for progress. *Hydrol. Process.* **2018**. [CrossRef]

61. Malvar, M.C.; Silva, F.C.; Prats, S.A.; Vieira, D.C.S.; Coelho, C.O.A.; Keizer, J.J. Short-term effects of post-fire salvage logging on runoff and soil erosion. *For. Ecol. Manag.* **2017**, *400*, 555–567. [CrossRef]

62. Santos, R.M.B.; Sanches Fernandes, L.F.; Pereira, M.G.; Cortes, R.M.V.; Pacheco, F.A.L. A framework model for investigating the export of phosphorus to surface waters in forested watersheds: Implications to management. *Sci. Total Environ.* **2015**, *536*, 295–305. [CrossRef] [PubMed]

63. Shakesby, R.A.; Boakes, D.J.; Coelho, C.O.A.; Gonçalves, A.J.B.; Walsh, R.P.D. Limiting the soil degradational impacts of wildfire in pine and eucalyptus forests in Portugal: a comparison of alternative post-fire management practices. *Appl. Geogr.* **1996**, *16*, 337–355. [CrossRef]

64. Li, Z.; Fang, H. Impacts of climate change on water erosion: A review. *Earth-Sci. Rev.* **2016**, *163*, 94–117. [CrossRef]

65. Paroissien, J.-B.; Darboux, F.; Couturier, A.; Devillers, B.; Mouillot, F.; Raclot, D.; Le Bissonnais, Y. A method for modeling the effects of climate and land use changes on erosion and sustainability of soil in a Mediterranean watershed (Languedoc, France). *J. Environ. Manag.* **2015**, *150*, 57–68. [CrossRef]

66. Smetanová, A.; Nunes, J.; Symenoakis, E.; Brevik, E.; Schindelwolf, M.; Ciampalini, R. Mapping and modelling soil erosion to address societal challenges in a changing world. *Land Degrad. Dev.* **2019**. [CrossRef]

67. De Figueiredo, T.; Fonseca, F.; Lima, E.; Fleischfresser, L.; Hernandez, Z. *Assessing Performance of Post-Fire Hillslope Erosion Control Measures Designed for Different Implementation Scenarios in NE Portugal: Simulations Applying USLE*; Nova Scien.: Bragança, Portugal, 2017.

68. Prats, S.A.; Wagenbrenner, J.W.; Martins, M.A.S.; Malvar, M.C.; Keizer, J.J. Mid-term and scaling effects of forest residue mulching on post-fire runoff and soil erosion. *Sci. Total Environ.* **2016**, *573*, 1242–1254. [CrossRef]

69. Prats, S.A.; dos Santos Martins, M.A.; Malvar, M.C.; Ben-Hur, M.; Keizer, J.J. Polyacrylamide application versus forest residue mulching for reducing post-fire runoff and soil erosion. *Sci. Total Environ.* **2014**, *468*, 464–474. [CrossRef]

70. Brauman, K.A.; Daily, G.C.; Duarte, T.K.; Mooney, H.A. The nature and value of ecosystem services: An overview highlighting hydrologic services. *Annu. Rev. Environ. Resour.* **2007**, *32*, 67–98. [CrossRef]

71. Jones, O.D.; Nyman, P.; Sheridan, G.J. A stochastic coverage model for erosion events caused by the intersection of burnt forest and convective thunderstorms. In Proceedings of the MODSIM 2011—19th International Congress on Modelling and Simulation—Sustaining Our Future: Understanding and Living with Uncertainty, Perth, Australia, 12–16 December 2011; pp. 2338–2344.

72. Lopes, A.F.; Macdonald, J.L.; Quinteiro, P.; Arroja, L.; Carvalho-Santos, C.; Cunha-e-Sá, M.A.; Dias, A.C. Surface vs. groundwater: The effect of forest cover on the costs of drinking water. *Water Resour. Econ.* **2019**, *28*, 100123. [CrossRef]

73. Mulumba, L.N.; Lal, R. Mulching effects on selected soil physical properties. *Soil Tillage Res.* **2008**, *98*, 106–111. [CrossRef]

Article

Multiple Temporal Scales Assessment in the Hydrological Response of Small Mediterranean-Climate Catchments

Josep Fortesa [1,2,*], Jérôme Latron [3], Julián García-Comendador [1,2], Miquel Tomàs-Burguera [4], Jaume Company [1,2], Aleix Calsamiglia [1,2] and Joan Estrany [1,2]

1   Mediterranean Ecogeomorphological and Hydrological Connectivity Research Team, Department of Geography, University of the Balearic Islands, Ctra. Valldemossa km 7.5, 07122 Palma, Balearic Islands, Spain; julian.garcia@uib.cat (J.G.-C.); jaume.company@uib.cat (J.C.); aleix.calsamiglia@gmail.com (A.C.); joan.estrany@uib.cat (J.E.)
2   Institute of Agro—Environmental and Water Economy Research—Inagea, University of the Balearic Islands, Ctra. Valldemossa km 7.5, 07122 Palma, Balearic Islands, Spain
3   Institute of Environmental Assessment and Water Research (IDAEA), Spanish Research Council (CSIC), Jordi Girona 18, 08034 Barcelona, Spain; jerome.latron@idaea.csic.es
4   Estación Experimental de Aula Dei, Consejo Superior de Investigaciones Científicas (EEAD-CSIC), 50192 Zaragoza, Spain; mtomas@eead.csic.es
*   Correspondence: josep.fortesa@uib.cat

Received: 13 November 2019; Accepted: 14 January 2020; Published: 19 January 2020

Abstract: Mediterranean-climate catchments are characterized by significant spatial and temporal hydrological variability caused by the interaction of natural as well human-induced abiotic and biotic factors. This study investigates the non-linearity of rainfall-runoff relationship at multiple temporal scales in representative small Mediterranean-climate catchments (i.e., <10 km^2) to achieve a better understanding of their hydrological response. The rainfall-runoff relationship was evaluated in 43 catchments at annual and event—203 events in 12 of these 43 catchments—scales. A linear rainfall-runoff relationship was observed at an annual scale, with a higher scatter in pervious (R^2: 0.47) than impervious catchments (R^2: 0.82). Larger scattering was observed at the event scale, although pervious lithology and agricultural land use promoted significant rainfall-runoff linear relations in winter and spring. These relationships were particularly analysed during five hydrological years in the Es Fangar catchment (3.35 km^2; Mallorca, Spain) as a temporal downscaling to assess the intra-annual variability, elucidating whether antecedent wetness conditions played a significant role in runoff generation. The assessment of rainfall-runoff relationships under contrasted lithology, land use and seasonality is a useful approach to improve the hydrological modelling of global change scenarios in small catchments where the linearity and non-linearity of the hydrological response—at multiple temporal scales—can inherently co-exist in Mediterranean-climate catchments.

Keywords: rainfall-runoff; multiple temporal scales; non-linearity; small catchments; Mediterranean

---

1. Introduction

The complexity of Mediterranean fluvial systems is caused by the multiple temporal and spatial heterogeneities in the relationships between the natural and human-induced abiotic and biotic variables, especially in the Mediterranean Sea Region [1,2]. The Mediterranean climate lies between 32° and 40° N and S of the Equator and is characterized by a wet and mild winter, a warm and dry summer and a high inter- and intra-annual variability in rainfall patterns. Mediterranean climate regions comprise the Mediterranean Sea Region, the coast of California, Central Chile, the Cape region of South Africa and the southwestern and southern parts of Australia [3]. Under these climatic conditions, catchments are mostly characterized by a high diversity in hydrological regimes [4,5] promoting significant temporal and spatial differences in the hydrological response [6–8]. The seasonality of the Mediterranean climate plays a key role in the runoff generation processes, increasing the non-linearity of the rainfall-runoff relationship at the event scale [9–11]. In winter and early spring, saturation processes are dominant, due to water reserves triggering the runoff generation [8,12]. During late spring, summer and early autumn, those same authors observed how runoff was generated under Hortonian conditions due to high rainfall intensities. Different runoff mechanisms can co-exist within a catchment [13], although flood events under antecedent saturation wetness conditions enable a larger hydrological response [12,14–16].

The spatial variability of runoff generation is also elucidated in the catchment hydrological response as a combination of rainfall-distribution [17] and runoff-contribution areas [18]. Thus, the spatial distribution of landscape elements in relation to each other is fundamental in influencing transfer flow pathways [19]. The main factors governing connectivity are associated to changes in topography [20], soil properties [21], soil type [22] and in vegetation cover [23]. In addition, river connectivity occurs along a three spatial dimension (i.e., longitudinal, lateral and vertical) which can change markedly over time, especially in ephemeral rivers [24]. Consequently, flow pathway activation depends on rainfall amount and intensity, as well as soil moisture antecedent conditions [25–27]. Lithology's effects on runoff response in Mediterranean Sea Region fluvial systems are conditioned by the presence of karst features, as the proportion of carbonate rocks is significantly higher than in other landscapes [28]. Therefore, its characteristics related to hydrology, such as high infiltration rates, deep percolation and spring sources, must be taken into account [29–31]. In limestone areas, Hortonian and saturation runoff can both be generated and infiltrated downslope [32,33], whilst in badland areas the runoff generation is characterized by a lower soil infiltration capacity than in karst areas [34]. In areas with high clay subsoil content [35] and over granite bedrock [36], runoff is generated as a combination of a lack of deep percolation and a subsurface runoff over an impervious subsoil layer, even artificially promoted [15].

Land uses also alter the hydrological response, depicting a runoff reduction when agriculture uses are replaced by forests [37]. Afforestation processes due to land abandonment can promote a 40% reduction in the annual water yield in Mediterranean Sea Region fluvial systems [38], whilst forest logging may increase the annual runoff coefficient by up to 16% [39]. However, afforested catchments generate the largest flows and peak discharges at the event scale when compared with forest catchments [40]. Most of these catchments are affected by soil and water conservation structures historically built to reduce overland flow and prevent erosion [41]. The abandonment and degradation of these structures may promote and increase runoff and sediment yield [42], with runoff coefficients being between 20% and 40% in abandoned terraces [43].

Hydrologists develop perceptual models of the catchments they study, which consist of an appreciation of the dominant processes controlling the hydrological response, using field measurements and observations [44]. Numerical models used to simulate catchment behaviour often fail build on this knowledge [45], also considering that the Mediterranean-climate catchments show very heterogeneous responses over time and space, resulting in limitations in hydrological modelling and large uncertainties in predictions [46]. To reduce this spatio-temporal scale variability, small experimental and representative catchments are useful to observe the hydrological response under different or specific land use, lithology and human effect characteristics [47]. The aim of this paper is to investigate the rainfall-runoff relationship at different temporal scales in representative small Mediterranean-climate catchments (i.e., <10 km^2), evaluating the role of lithology and land use. At the annual scale, the runoff response was assessed at 43 catchments under a pervious or impervious lithology. At the event scale, the rainfall-runoff response of 203 events was investigated to examine the effects of seasonality, lithology and land use. In addition, the inter- and intra-annual variability of the rainfall-runoff and the temporal downscaling (i.e., annual to event scale) was studied in the Es Fangar Creek catchment (3.35 km^2; Mallorca, Spain) during five hydrological years. The rainfall-runoff relationship assessment under contrasted lithology, land use and seasonality may provide new insights into the hydrological modelling of small Mediterranean-climate catchments. These catchments are highly demanding in terms of data and event-scale modelling because classical hydrological models are not well suited to the Mediterranean area, as many hydrological processes remain poorly represented [46].

2. Materials and Methods

2.1. Study Areas

2.1.1. Small Mediterranean-Climate Catchments

A total of 43 small catchments (i.e., <10 km^2) from 22 published studies on the Mediterranean climate regions (Figure 1a) were selected to analyse hydrological response. The geographical distribution of the catchments was grouped into the main climate regions, as follows: (1) Western coast of USA, (2) Western Mediterranean Sea Region (from Spain to Italy), (3) Eastern Mediterranean Sea Region (Israel) and (4) South Africa. The area of the catchments ranged from 0.05 to 9.61 km^2, the median value being 1.03 km^2 and the standard deviation 2.6 km^2. The mean annual rainfall ranged from 367 to 1794 mm y^{-1}, with a median value of 833 mm y^{-1} $\pm$ 334 mm y^{-1}. The mean annual temperature ranged from 6.6 to 17.2 °C with a median value of 13.9 °C $\pm$ 3 °C. When temperature information was not available, it was obtained according to Fick and Hijmans [48]. The predominant lithology was pervious in 12 catchments, and was impervious in the other 31. Within the 43 catchments, the studies of 12 of them also contained information related to the main land uses, which was used in this paper for assessing their hydrological response at the event scale. The main land uses were agriculture (3 catchments), agroforestry (3), forestry (1) and shrub (5).

2.1.2. Es Fangar Creek

A temporal downscaling assessment of the inter- and intra-annual variability of the rainfall-runoff relationship was carried out in the Es Fangar Creek catchment, a headwater tributary of the Sant Miquel River catchment (151 km^2) located in the north-eastern part of Mallorca Island (Figure 1b) and is representative of Mediterranean mid-mountainous catchments. The lithology is mainly composed of marl and marl-limestone formations from the medium–upper Jurassic and Cretaceous period in the valley bottoms. In the upper parts of the catchment, massive calcareous and dolomite materials from the lower Jurassic period and dolomite and marl formations from the Triassic period (Rhaetian) are dominant. The Es Fangar catchment has an area of 3.4 km^2, with altitudes ranging from 72 m.a.s.l. to 404 m.a.s.l. (Figure 1c). The mean slope of the catchment is 26% and the length of the main channel is 3.1 km (average slope of 22%). The drainage network is natural in the headwater parts. In the bottom valley, flow lamination is applied, with transverse walls and also the straightening and diverting of the main stream, with the banks fixed with dry-stone walls for flood control and erosion prevention. In addition, subsurface tile drains are also installed to facilitate drainage due to the impervious materials which would impede agricultural activity during wet periods. As a result, 16% of the surface catchment is occupied by soil and water conservation structures. Since 1950, important socio-economic changes have caused a gradual abandonment of farmland in marginal areas, leading to afforestation. The land uses in 1956 were rainfed herbaceous crops (54%), forest (31%) and scrubland (15%). Nowadays, the main land uses (Figure 1c) are forest (63%), rainfed herbaceous crops (32%) and scrubland (5%). In addition, 54% of terraced land is currently covered by forests (Figure 1c), demonstrating the consolidation of the forest transition. The climate of the area is classified on the Emberger scale [49] as Mediterranean temperate sub-humid. The mean annual rainfall (1965–2016, Biniatró AEMET station) is 927 mm y^{-1} with a variation coefficient of 23%, and the mean annual temperature is 15.7 °C. A rainfall amount of 180 mm in 24 h is estimated to have a recurrence period of 25 years [50]. The Es Fangar streamflow regime can be classified as intermittent flashy (49% zero days flow), with an annual variability from intermittent (35% zero days flow) to harsh intermittent (62% zero days flow).

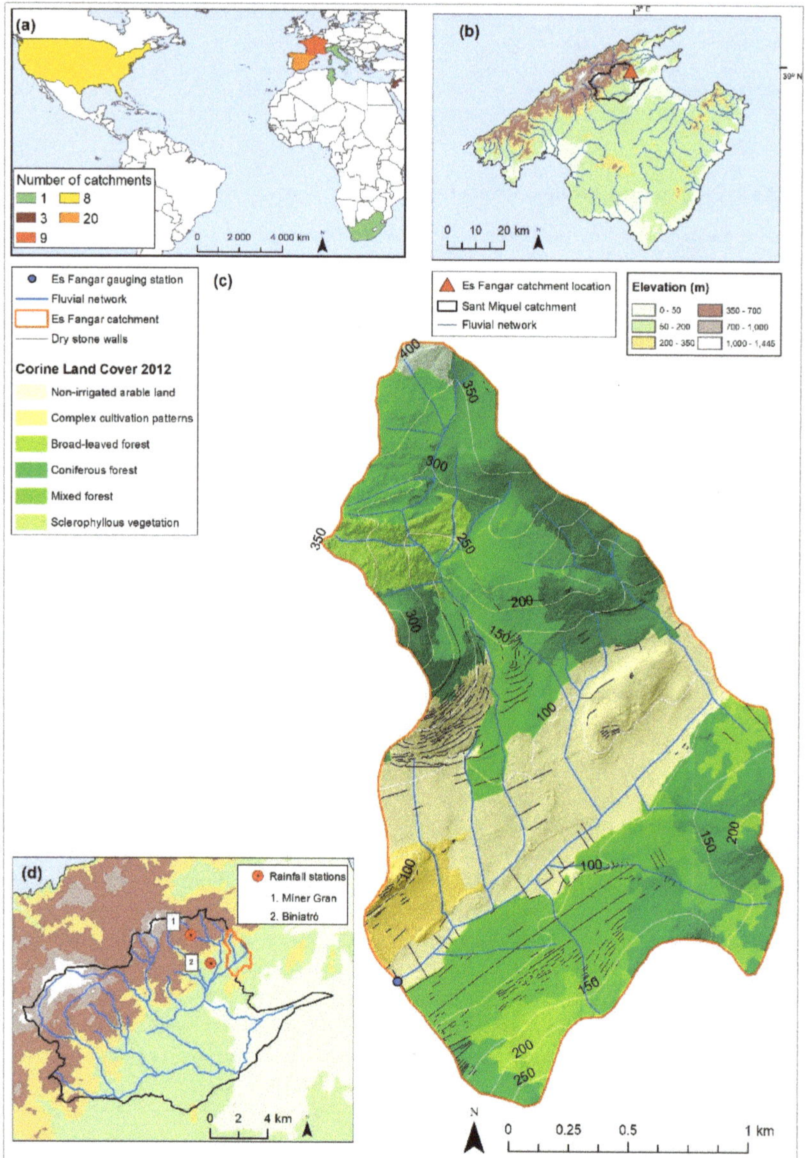

Figure 1. (**a**) Map of the small Mediterranean-climate catchments selected to assess the rainfall-runoff relationship at the annual and event scale. (**b**) Map of Mallorca Island, showing the location of the Sant Miquel River and Es Fangar Creek catchments. (**c**) Map of the Es Fangar Creek catchment, showing the different land-uses, the stream network and the gauging station. (**d**) Map of the Sant Miquel River catchment with the location of rainfall stations used in this study.

2.2. Hydrological Response of Small Mediterranean-Climate Catchments

Bivariate statistical regressions were used to establish the correlations at the annual and event scales between rainfall and runoff in order to assess the hydrological response of small Mediterranean-climate catchments (i.e., <10 km^2; Figure 1c). These equations were used descriptively as least squares adjusted for the rainfall-runoff assessment. At the annual scale, data from the 43 representative catchments (Table A1 (a); hereinafter A1a) were collected to observe the influence of lithology on this response; i.e., the catchments were classified as pervious or impervious by using the information regarding the catchments' characteristics (e.g., soil type, soil texture or lithology materials) extracted from research papers. At the event scale, 203 events from 12 representative catchments were classified according to (a) seasonal occurrence (autumn, winter, spring or summer), (b) pervious or impervious lithology and (c) main land use (agricultural, agroforestry, forest or shrub) (Table A1 (b)). The main land uses of each catchment were divided into specific land uses when the information was available (Figure 1). Most of these studies (79%) were located in the Mediterranean Sea Region and 47% of them studied catchments < 1 km^2.

2.3. Monitoring and Data Acquisition in Es Fangar

The rainfall data since 2012 were obtained from the B696 Biniatró AEMET station (Figure 1d; 1 km away from the catchment). In October 2014, a rainfall gauge station (Míner Gran) was installed less than 2.5 km away from the catchment. The Míner Gran rainfall gauge is located 1 m above the ground and connected to a HOBO Pendant$^{®}$ G Data Logger-UA-004-64 that records precipitation at 0.2 mm resolution. A linear regression was established (n = 978; R^2: 0.88) for daily rainfall (2014–2017) between the Biniatró and Míner Gran stations to reconstruct rainfall data series from 2012 to 2014 for the Míner Gran station. Due to the lack of temperature data available in the studied catchment, the data of neighbouring AEMET weather stations (i.e., less than 8 km away from the catchment) were used to estimate the catchment's temperature by using the block kriging technique. With this information, the monthly evapotranspiration (i.e., ET$_o$) was estimated using the equation from Hargreaves and Samani [51].

The gauging station of the Es Fangar catchment was built in July 2012. Its cross section is formed by a rectangular broad-crested weir for low water stages to better measure low discharges. The water level was continuously (1 min time step) measured using a pressure sensor Campbell CS451 connected to a CR200X datalogger and average readings were kept every 15 min. The flow velocity was measured during baseflow conditions and flood events using an OTT MF Pro electromagnetic water flow meter. These flow velocity measurements (n = 17) were subsequently used to calibrate the stage–discharge relationship.

2.4. Rainfall-Runoff Relationship Assessment in Es Fangar

This study is based on data from 5 hydrological years (2012–2013 to 2016–2017). For each runoff event (i.e., when the water stage exceeds the low-flows channel; i.e., 0.036 m^3 s^{-1}), a simple hydrograph separation between quickflow and baseflow components was performed through a visual technique based on the breakpoints detected on the logarithmic falling limb of the hydrograph [52]. This hydrograph separation method was used only to characterize the response of the catchment to a rainfall, with no aim of deriving any interpretation in terms of runoff processes. Over a 5-year period, hydrograph separation was conducted on 49 events. Based on the hydrograph separations, the baseflow and quickflow contributions for each hydrological year (October to September) were calculated.

Several variables were derived from the hyetograph and hydrograph from each rainfall-runoff event (Table 1). These variables aimed to characterize the pre-event conditions (antecedent precipitation in 24 h, AP1d; baseflow at the start of the event, Q$_0$) as well as the event characteristics (rainfall depth, P$_{tot}$; mean and maximum rainfall intensity, IP$_{mean}$30 and IP$_{max}$30; runoff depth and runoff coefficient, R and R$_c$; and peak-flow discharge, Q$_{max}$). The relationships between these variables were assessed through the Pearson correlation matrix. A more detailed analysis of rainfall-runoff relationships in the Es Fangar catchment was carried out in a second step to investigate the variability of the hydrological response to similar size rainfall events to assess the rainfall-runoff non-linearity. The widest range of events with a similar P_{tot} (40–50 mm) generating the highest differences in R response was selected. Accordingly, 7 events with a P_{tot} range from 41.8 to 49.8 mm but with different antecedent conditions or rainfall dynamics were selected and also classified in relation to wet, dry and transition periods, in accordance with Gallart et al. [53].

Table 1. Pre-event and event conditions variables to explain the rainfall-runoff relationship in the Es Fangar Creek catchment.

Pre-Event Conditions		Event Conditions	
Q_0	Baseflow at the start of the flood (m^3 s^{-1})	P_{tot}	Rainfall depth (mm)
AP1d	Antecedent precipitation 1 day before (mm)	$IP_{mean}30$	Average rainfall intensity (mm h^{-1})
		$IP_{max}30$	Maximum 30′ rainfall intensity (mm h^{-1})
		Q_{max}	Maximum peak discharge (m^3 s^{-1})
		R	Runoff (mm)
		R_c	Runoff coefficient

3. Results

3.1. Hydrological Response of Small Mediterranean-Climate Catchments

3.1.1. Annual Scale: Lithology Influence

The annual rainfall ranged from 376 to 1794 mm yr^{-1}, whilst R ranged from 3.7 to 1200 mm within these representative catchments under Mediterranean-climate conditions. The relationship between annual rainfall and R (Figure 2) showed a significant positive linear correlation ($R^2 = 0.68$; $p < 0.01$). However, some scattering was also apparent in the relationship because when catchments with pervious lithology were not included, the regression increased ($R^2 = 0.82$; $p < 0.01$; see Figure 2). The lowest annual R values (<10 mm) are related to catchments with annual rainfall < 500 mm yr^{-1} and a pervious lithology (Figure 2), located in the Eastern Mediterranean Sea (Lower Jordan River). Besides, catchments with pervious lithology with ca. 1500 mm yr^{-1} of annual rainfall showed large differences in their R amount, ranging from 70 to 974 mm.

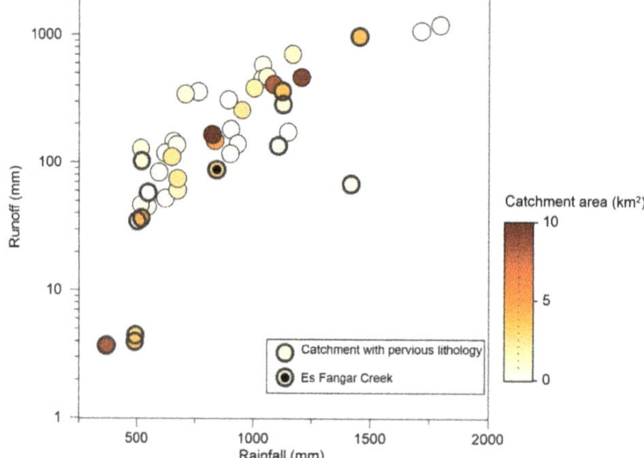

Figure 2. Rainfall-runoff for 43 small Mediterranean-climate catchments at the annual scale. Catchments with pervious lithology are marked with a grey halo. The Es Fangar Creek value is illustrated with a black dot.

3.1.2. Event Scale: Seasonality, Lithology and Land Use Influences

To investigate the variability of the hydrological response at the event scale, the rainfall-runoff relationships of 203 events from 12 small Mediterranean-climate catchments were analysed (Table A1 (b) and Figure 3). Within Figure 3, the outliers are marked with an ellipsoid and excluded from the correlations because outliers enlarged the rating curves of these correlations, obtaining significant relationships in all cases.

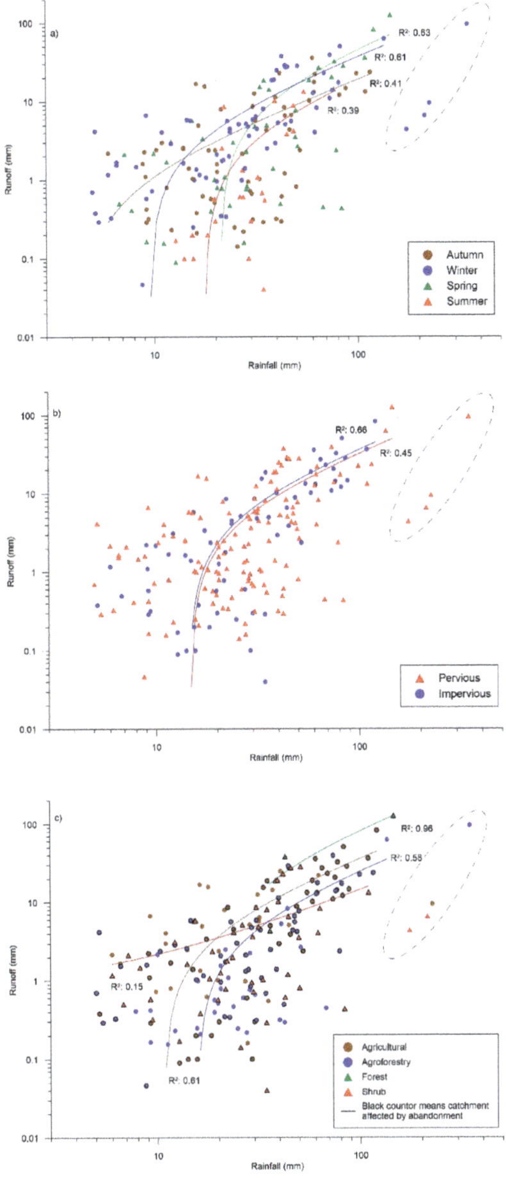

Figure 3. Rainfall-runoff relationship at the event scale classified by (**a**) season, (**b**) lithology and (**c**) land uses at 12 small Mediterranean-climate catchments. Outliers are marked with an ellipsoid.

The seasonal distribution of the events was winter (34%), autumn (32%), spring (22%) and summer (12%)—most of them occurred between November and February (47%). Similar seasonal median rainfall was observed, ranging from 26.2 (winter) to 29.0 mm (autumn). The highest seasonal median of event R occurred in winter (4.2 mm). However, those events occurred during the transition periods depicted a similar median R (autumn = 2.3 mm; spring = 2.2 mm) being the lowest value for summer events (0.6 mm). In the case of rainfall-runoff relationships, the highest correlation was obtained in spring (R^2 = 0.63), followed by winter, autumn and summer (Figure 3a).

Small differences between the median of R events in catchments with pervious lithology (2.2 mm) and catchments without pervious lithology (3.1 mm) were observed. Nevertheless, significant differences in rainfall-runoff relationships were detected as events in catchments with pervious lithology showed the highest correlation and the lowest scattering (Figure 3b). Rainfall events > 55 mm generated a R > 4 mm, excepting events occurred in late spring (Figure 3a) in catchments with predominance of pervious lithology (Figure 3b).

Event rainfall-runoff relationship was carried out considering the differences in the main land uses of studied catchments (Figure 3c). The median event R in forest (37.6 mm) was higher than agricultural (5.0 mm), shrub (2.0 mm) and agroforest (1.5 mm) catchments. However, the highest correlations between rainfall and R were obtained in catchments with a predominance of forest (R^2 = 0.96), agricultural (R^2 = 0.61) and agroforest (R^2 = 0.58) land uses.

3.2. Hydrological Response at Multiple Temporal Scales in Es Fangar Catchment

3.2.1. Annual Scale

The mean annual rainfall (840.8 mm yr^{-1} ± 213 mm y^{-1}) calculated over the five hydrological years (2012–2017) was broadly representative (−9%) of the long term mean annual rainfall (927 mm y^{-1} ± 215 mm y^{-1}: 1965–2016), but showed a high range of variation (coefficient of variation 25%, being 553 mm y^{-1} to 1090 mm y^{-1}), characteristic of Mediterranean conditions. Figure 4 shows the annual variability of rainfall, baseflow and quickflow contributions as well the annual R_c and the number of days with observed flow at the gauging station. Linear positive relationships were observed between annual rainfall, R, R_c, baseflow and quickflow ($R^2 \geq 0.84$, data not shown). The annual R_c ranged from 2.9% to 14.2% (mean value = 10.4%) and quickflow contribution from 9.9% to 45.0% (mean value = 33.0%). During the study period, flow was observed 42.8% of the time. From year to year, the cumulated number of days with flow ranged from 37.8% (138 days) to 65.2% (238 days) of the total days (Figure 4). An inverse relation between annual R and the number of days with flow was established. On the one hand, hydrological years with the largest R (2013–2014, 2014–2015 and 2016–2017) showed fewer days with flow and a lower baseflow contribution (<60%). In these years, 50% to 62% of the annual R was reached in 5 or fewer days (Table 2). In addition, days with more R were always during autumn and winter. On the other hand, the hydrological years with the lowest R (2012–2013 and 2015–2016) showed 208 and 238 days with flow. The baseflow contributions represented 70% and 90%, respectively, and 50% of the annual R was reached in 10 and 17 days, also respectively. The contribution of the 5 days with more R did not exceed 28% and 37%, respectively, and these days with more R were distributed among the four seasons.

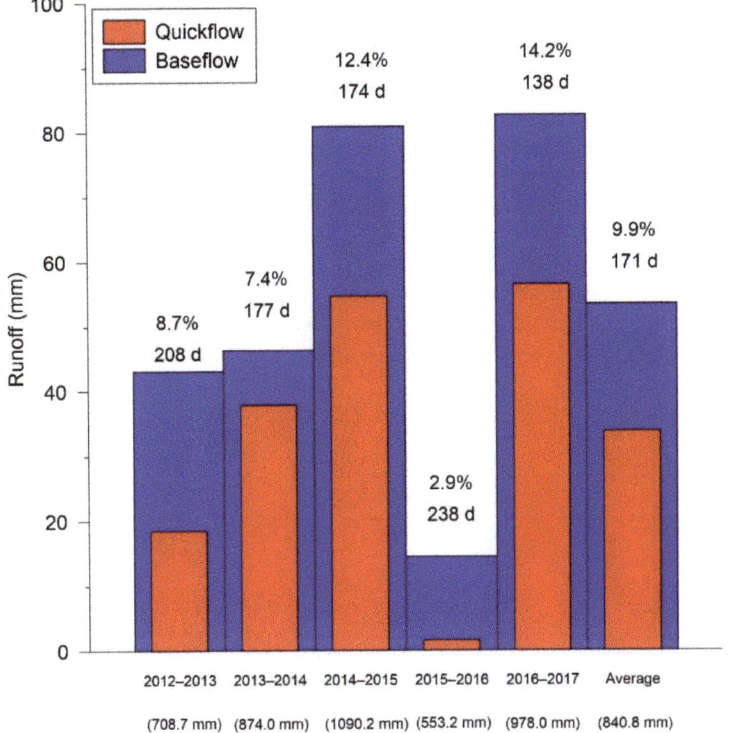

Figure 4. Baseflow and quickflow contributions (mm) in the total flow for each hydrological year (October to September) at Es Fangar Creek. The annual R_c is depicted in %, as well as the total number of days (d) with recorded flow at the gauging station. The total annual rainfall is also illustrated in brackets.

3.2.2. Seasonal Scale

Figure 5 shows the monthly mean values of rainfall, reference evapotranspiration and minimum, median and maximum R values. The mean monthly rainfall values were broadly comparable to those observed for the long-term period, except for October, when rainfall during the study period was much lower (56% lower than long-term rainfall).

The seasonal dynamics of rainfall and evapotranspiration controlled the R response. Characteristic wet (winter) and dry (summer) periods alternated throughout the year, separated by transition periods (last autumn and early spring). Autumn was the rainiest season, with low evapotranspiration and flow observed at the outlet for 52.7% of the time, on average. The autumn mean R_c was 9.1% and ranged from 4.1% to 11.4%. During November and December, a continuous baseflow was generated in response to a large rainfall amount after the pre-filling of the initial water storage in October. From October to December, the mean R_c increased from 1.3% to 13.1%. Finally, the minimum, median, average and maximum monthly R amounts in December were higher than in November, even with lower rainfall amounts (Figure 5). Although differences between the hydrological years were observed, autumn was the season with less inter-variability in terms of R response.

During winter, flow occurred during 90.6% of the time, although the mean rainfall amount (294 mm) was lower than in autumn (326 mm). This more frequent presence of flow was caused by the low evapotranspiration demand, thus keeping the hydrological pathways active, as already established in autumn. Consequently, in winter, a R amount similar to that of autumn can be generated from a lower rainfall amount. The winter mean R coefficient was 16.9%, ranging from 1.5% to 27%. During winter, the mean monthly R_c decreased from 18.9% to 14.6% between January and March as a consequence of a decreasing rainfall amount and an increasing evapotranspiration demand.

In spring, flow was observed 41.3% of the time. The monthly rainfall decreased during the season and was much lower (127.4 mm) than in autumn and winter. Although monthly evapotranspiration losses increased and were higher than rainfall, the water reserves accumulated during the autumn and winter months sustained the flow contribution. The spring mean R_c was 5.1%, ranging from 0.4% to 8.4% during the study period. The average monthly R decreased during the season from April (7.7%) to June (0.3%).

Finally, in summer, flow was observed only 0.9% of the time. The stream was most often dry in this season due to the high negative balance between rainfall and evapotranspiration. R was only ephemeral in response to the episodic rainfall events more frequent in September, when 80% of rainfall occurred on average (Figure 5). The summer R_c was 1.4%.

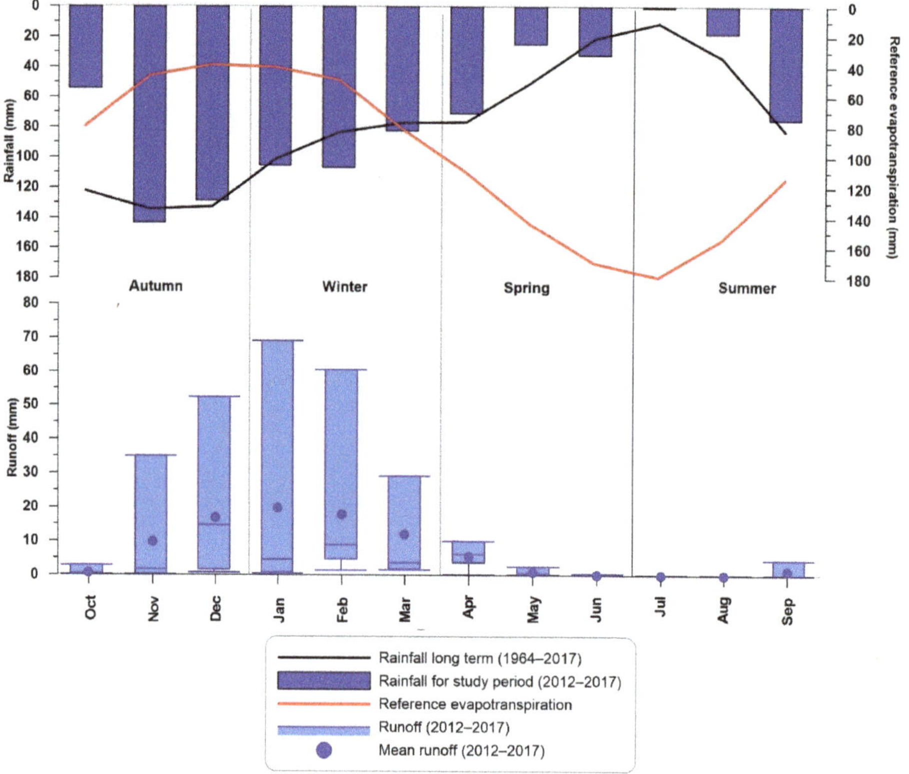

Figure 5. Monthly time series of rainfall, R, reference evapotranspiration during the study period (2012–2017) at Es Fangar Creek. Box plots show minimum, median and maximum monthly R. Blue dots show mean monthly R. Long term (1964–2017) monthly rainfall distribution is also depicted.

3.2.3. Non-Linearity Assessment at Event Scale

During the study period, the number of flood events per hydrological year was between 2 (2015–2016) and 14 (2014–2015). Their seasonal distribution was 13 events in autumn, 25 in winter, 7 in spring and 4 in summer. The events' characteristics are detailed in Table 3. An investigation of the variability and the non-linearity of the hydrological response at the event scale was carried out by means of a comparative assessment of seven rainfall-runoff events with similar P_{tot}, ranging from 41.8 to 49.8 mm, but different antecedent conditions and rainfall dynamics (Figure 6).

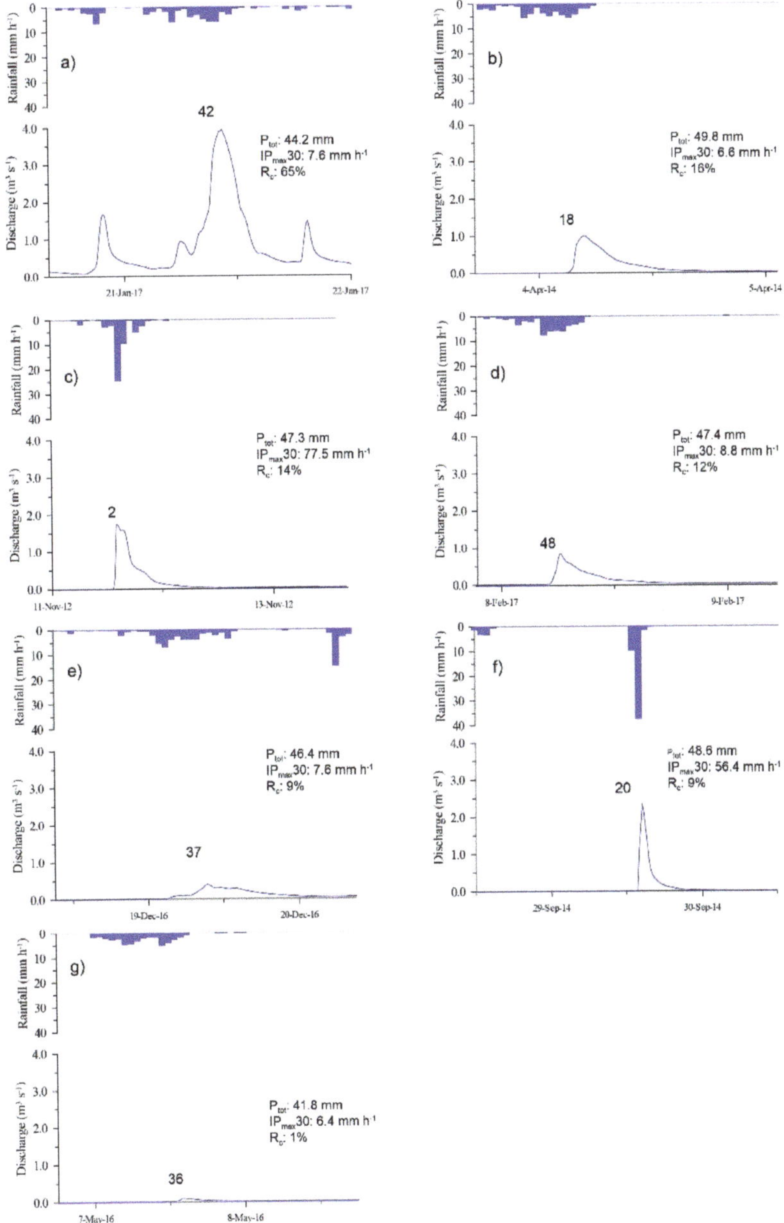

Figure 6. Selected events for non-linearity analysis at Es Fangar Creek. Events with a total precipitation between ca. 40 and 50 mm. The numbers located at the peak of each hydrograph indicate the ID of the events recorded in Es Fangar Creek during the study period 2012–2017, further exposed in Table 3.

The event on 21 January 2017 (Figure 6a) occurred under wet conditions following a runoff event on the day before. AP1d was 35.4 mm and Q_0 was high (0.229 m^3 s^{-1}). The rainfall characteristics (duration and intensity) were similar to those observed for other events of the same magnitude (Figure 6b,e,g), but the hydrological response was an order of magnitude higher in terms of R (28.9 mm), R_c (65%) and Q_{max} (3.942 m^3 s^{-1}) due to wet antecedent conditions (Table 4). This event had a similar duration to events that occurred under higher rainfall intensities (11 November 2012, 29 September 2014). However, wet antecedent conditions were much more favourable than rainfall intensity to reach a high Q_{max} and especially a higher R contribution.

The event on 4 April 2012 (Figure 6b) occurred in early spring, ergo at the beginning of the transition period from wet to drier conditions. The antecedent conditions were favourable to R generation, due to the presence of a baseflow maintained by water reserves accumulated in the autumn and winter months. R_c was 16%. The event on 11 November 2012 (Figure 6c) recorded the highest $IP_{max}30$ (i.e., 77.5 mm h^{-1}) of the study period, occurring in the transition period from dry to wet conditions (middle autumn). Previous to this event, the stream channel was dry. Under these conditions of rainfall intensity and without previous baseflow, the R_c reached 14%. Although the event on 4 April 2012 recorded a similar R_c (16%) to the 11 November 2012 event, both events resulted in different hydrographs because of different rainfall characteristics (in terms of duration and intensity) and antecedent wetness conditions. As a result, the duration of 44.7 h for the event on 4 April 2012 contrasted with the duration of the 11th November 2012 event, which only lasted 18 h. In addition, the differences between these two events were also relevant in terms of Q_{max}—1.006 m^3 s^{-1} for the April event and 1.747 m^3 s^{-1} for the event on 11 November 2012.

The event on 8 February 2017 (Figure 6d) occurred during the wet season. AP1d and a Q_0 were, respectively, 0 mm and 0.012 m^3 s^{-1}. On the contrary, the event on 19 December 2016 (Figure 6e) occurred at the end of the transition period and had an AP1d and a Q_0 of 2 mm and 0.012 m^3 s^{-1}, respectively. Both events had similar values of $IP_{max}30$, AP1d and Q_0 (Table 4). Nevertheless, R and R_c were slightly larger for the event on 8 February 2017, likely due to the large rainfall amount accumulated during the two events (more than 400 mm of rainfall between 19 December 2016 and 8 February 2017).

The event on 7 May 2016 (Figure 6g) was similar to the event on 19 December 2016 (Figure 6e) in terms of P_{tot}, duration and intensity. Nevertheless, R response started earlier (i.e., shorter response time) for the event on 19 December 2016. This difference was caused because the December 2016 event occurred during the wetting-up period when catchment water reserves were increasing. As a result, whilst the R response was relatively small (4.4 mm), R_c (9%) and Q_{max} (0.394 m^3 s^{-1}) were significantly larger than for the event on 7 May 2016 (Table 4), which occurred during the transition period towards dry conditions. With an AP1d of 1.6 mm and Q_0 of 0.006 m^3 s^{-1}, the lowest R response was generated (see total runoff, R_c and Q_{max} in Table 4). This R response was also the most delayed, as shown on Figure 6g, starting when 95% of the rain (39.8 mm) had fallen on the catchment.

The event on 29 September 2014 (Figure 6f) occurred during the dry season (i.e., late summer, early autumn) but with an AP1d of 11.8 mm. The rainfall event was the shortest of these seven selected events (3.2 h), with the second highest $IP_{max}30$ (56.4 mm h^{-1}) of the study period. This very high intensity triggered a R response during dry conditions. Similar rainfall but lower intensities during dry conditions always produced lower responses ($R_c \leq 3\%$). Although the R response was relatively small (4.2 mm) with $R_c < 10\%$, Q_{max} was relatively high (2.356 m^3 s^{-1}), promoted by the rainfall intensity.

4. Discussion on Hydrological Responses in Small Mediterranean-Climate and Es Fangar Catchments

4.1. Annual and Seasonal Scales

4.1.1. Small Mediterranean-Climate Catchments: Lithology Influences

Catchments with annual rainfall ranging from 500 to 900 mm yr^{-1} showed annual R ca. 20–350 mm, where the role played by lithology on the annual R response was important. Accordingly, the highest annual R values (>350 mm) were observed in catchments with badland areas [34] and a high subsoil clay content promoting lateral flow and a perennial regime [35]. Swarowsky et al. [54] also reported an annual R 17% lower than in an adjacent catchment to that investigated by Lewis et al. [35] and related this difference to greater deep infiltration. Similarly to our results, Merheb et al. [46] obtained a significant rainfall-runoff correlation (R^2 = 0.69) using 160 catchments from the Mediterranean Region (0.35 to 21,700 km^2). These authors highlighted that the scattering observed was due to karstic catchments or snowmelt contributions.

Also, scattering can be observed in the relationship for catchments with annual rainfall >1000 mm yr^{-1}. For this group of catchments, the annual R ranged from 135 to 1200 mm, except for the Santa Magdalena catchment, which is entirely on a limestone lithology and yielded an annual R value of 70 mm [47]. Large annual R values usually corresponded to watertight catchments with R values > 1000 mm yr^{-1} for two small Mediterranean-climate catchments on granite bedrock [36]. However, karstic environments can also promote very high annual R values, e.g., Ref. [55], as a result of incoming karstic spring sources. The conceptual model developed by Borg Galea et al. [2], which is based in exogenous and endogenous variables that link natural and human-derived variables and their influences in the system on a spatial–temporal scale, emphasized that climate is the main exogenous driver of the rainfall-runoff relationship, although their processes through precipitation, temperature and wind are mediated by catchment geology (endogenous variable).

Different seasonal rainfall-runoff relationships in relation to the influence of rainfall and evapotranspiration [56] are related to the alternation of rainfall and reference evapotranspiration throughout the year reproducing wet (winter), dry (summer) and transition periods (last autumn and early spring) [57]. Accordingly, winter is the season with the highest R_c, from 17% to 56%, due to rainfall being accumulated during autumn—and maintained in winter—and low evapotranspiration demand [10,15,56]. The same pattern was also observed during the five hydrological years assessed in Es Fangar Creek. In addition, all these studies also emphasized that in summer the hydrological response is limited or null, with R_c < 10%. Serrano-Muela [56] observed runoff coefficients < 5% in a range of rainfall amounts of 15–200 mm during summer in a small catchment characterized by a large forest cover. The driest environments in these studies obtained a summer R_c < 2%. R_c in autumn and spring were similar, ranging from 9% to 25% and 5% to 28% respectively. However, a variability throughout the seasons can be observed, especially in late autumn and early spring, as these are transition periods. Additionally, Lana-Renault [10] observed that the wetting-up period was steeper than the drying-down period, which was more progressive in a catchment located in the Pyrenees. Autumn was the season after the driest period, and it was when the evapotranspiration reached the maximum values. The beginning and the end of this season can be quite different in relation to R_c. The catchments have a null or limited response until a succession of rainfall events that fill the water reserves and generate favourable conditions for R generation [56]. The findings of Lana-Renault [10] in spring concluded that the accumulated rainfall during autumn and winter maintain high water reserves, allowing high runoff coefficients, even though spring was not the rainiest season.

4.1.2. Es Fangar

In the Es Fangar catchment, a low mean annual R value (87 mm) was recorded during the 2012–2017 period, despite its mean annual rainfall value (black dot on Figure 2). This result is most likely related to the presence of massive karstic dolomites and breccias in the lithology of the catchment headwaters.

The reported annual R in Es Fangar is within the R range of small catchments located in Mallorca, with annual rainfall ranging from 500 to 900 mm yr^{-1}, which yielded annual R between 36 and 130 mm. From these catchments, the highest R value was observed in an agricultural small lowland catchment prone to receiving subsurface contributions [15]. Spatial differences in the annual R between the headwater (102 mm) and catchment outlet (36 mm) due to transmission losses along the main channel were detected in the Sa Font de la Vila River, a mid-mountainous terraced catchment affected by forest fires [58]. Within this catchment, temporal differences were observed as R_c ranged from 1% to 22% from one year to year.

4.2. Event Scale

4.2.1. Seasonality, Lithology and Land Use Influences on Small Mediterranean-Climate Catchments Hydrological Response

The seasonal dynamics of rainfall and evapotranspiration during the year generate wet, dry and transition periods, which control the R response [10,53]. Therefore, seasonally, the highest correlation observed in spring could be related to the water reserves accumulated during the autumn and winter seasons, which is favourable for R generation [13]. According to the seasonal distribution of the 203 events, other studies in Mediterranean-climate catchments observed an event seasonality in the hydrological response [8,10,59] where autumn rainfall events produced a limited or null R response, especially at the beginning of the season. In addition, in winter and spring, the R_c values were always larger than 3%, being the maximum in spring (70%).

The role of transmission losses due to lithology has been investigated [32,60], depicting how low runoff values were recorded due to infiltration. In this study (see Section 3.2), events assessed in catchments with a predominance of pervious lithology showed a higher scattering than events in impervious catchments. In addition, rainfall events >55 mm generated R > 4 mm, except in events occuring in late spring in catchments with pervious lithology. In this case, the transmission losses due to pervious lithology have more influence on the hydrological response than the seasonal role of the water catchment reserves. Ries et al. [61] identified a low and large R response in rainfall events >50 mm based on the physiographic catchment and rainfall characteristics. Events occurring in catchments, which even had different land uses, with less bedrock permeability and less soil water storage but with higher values of rainfall intensity generate a greater R response than catchments with more bedrock permeability and soil water storage under low rainfall intensities. Contrarily, other types of lithology (i.e., badland areas) are characterized by a low infiltration capacity [62]. Latron and Gallart [63] observed that frequently badlands are the only active area to R response. According to these findings, Nadal-Romero et al. [64] observed an event stormflow coefficient of up to 20% in a Mediterranean-climate catchment with badland areas, even when the headwater catchment was forested (30% of the catchment area).

Events occurring in agricultural catchments obtained the second highest values in the median R and in the rainfall-runoff correlation, as arable land and its spatial distribution within a catchment have been demonstrated to be a driver for R generation [65]. However, events in the forest catchment had the highest correlation and median R, as the water yield and rainfall-runoff relationship decrease in forests catchments when compared to agricultural catchments, due to forest cover [66]. However, this correlation (R^2: 0.96; n: 3) and median value is related to the presence of karstic springs in the forest catchment, which was the only forest catchment found in the literature. The median R of forest catchments could be similar to the values obtained in afforested ones, as both land uses trigger a reduction in the R generation. This reduction was quantified in a loss of water yield ca. 10–40% [38,66–69]. Nevertheless, larger flows and Q_{max} were observed in afforested catchments than forest ones. The shrub median R (2.0 mm) was higher than afforested catchments, as shrub coverage is characterized by lower vegetation cover than forest and afforested land uses. In addition, the second main land use in shrub catchments is agricultural land use (Figure 1A), which could explain the higher median R compared with the afforested catchments. The same results were observed in García-Ruiz et al. [66], where catchments under shrub land use had lower R contributions than catchments with forest land use due to a lower plant cover density in the shrub catchment.

4.2.2. Rainfall-Runoff Relationship at the Es Fangar Catchment

The identification of driving factors explaining the variability of R_c can be explored through the relationships between R_c and R and P_{tot}, rainfall, maximum intensity and baseflow at the beginning of the event, as Latron et al. [8] carried out in a mountainous Mediterranean-climate catchment in North Eastern Spain.

Only the correlation between R and R_c was significant in the Es Fangar catchment (Figure 2a), explaining 73% of the variance. As in Latron et al. [8] this relationship had the highest correlation. However, contrarily to Latron et al. [8], the relationship between rainfall and R_c was highly non-lineal in Es Fangar, presenting a huge scattering, especially for P_{tot} between 20 and 60 mm (Figure 2b), which yielded R_c ranging from 1% to 65%. For larger rainfall events (i.e., 60 to 100 mm), the R_c values were high (>20%), but low R responses (with R_c ranging 1% to 3%) were also observed when large rainfall events were recorded after or during dry conditions. In line with the results from Latron et al. [8], R_c was not related with rainfall intensity (Figure 2c).

The relationship between Q_0 and R_c was not linear in the Es Fangar catchment (Figure 2d). A threshold was identified leading to larger R_c (>10%) for events with Q_0 above 0.04 m^3 s^{-1}, which always occurred between November and April. The relationship between Q_0 and R_c in Latron et al. [8] was stronger, because of the presence of a baseflow during most of the year. In general terms, the frequent absence of a baseflow in the Es Fangar catchment between May and October worsens most of the relationships presented in Figure 2 in comparison to a catchment with an almost permanent baseflow, as in Latron et al. [8] and Nadal-Romero et al. [40]. Then, baseflow–R_c correlation increased as the environment got wetter, from $R^2 = 0.19$ in Es Fangar to $R^2 = 0.70$ in a humid catchment [70].

In order to help in identifying the main factors involved in the hydrological response in the Es Fangar catchment, multiples relationships were carried out to investigate the influence of rainfall, Q_0 and $IP_{max}30$ on R and Q_{max} (Figure 7). The hydrological response by using R and Q_{max} was highly non-linear, as observed in other Mediterranean-climate catchments [8,11] because the soil moisture antecedent conditions are crucial for R generation. Figure 7a shows how the R response could not be explained by Q_0, as the highest R values presented a large scattering along the Y axis. This pattern confirmed that large R amounts (i.e., >10 mm) may occur from large rainfall events occurring in dry or wet conditions (i.e., with low or high Q_0 values). Therefore, P_{tot} was a key factor for R response, as shown by the significant positive correlation (p < 0.01) between both variables (Table 2). The largest R values were always recorded during last autumn or winter, with contributions of >17 mm (Figure 3 and Table 5), and are characterized by P_{tot}, R, R_c, Q_{max} and AP1d values much higher than median values (Table 3). Contrarily, large rainfall events occurring in spring resulted in a small R contribution because of the dry antecedent conditions (i.e., low or null Q_0 values). Other authors also assessed relationships between seasonality and R generation. In a mountainous catchment, Lana-Renault [10] observed that the highest rainfall-runoff correlations at the event scale were depicted in winter and spring due to higher water reserves than in autumn and summer. In Es Fangar, a group of events during autumn and winter characterized by $Q_0 > 0.080$ m^3 s^{-1} always generated R_c greater than 14% (Figure 2d). The P_{tot} of these events ranged from 5.2 to 61.4 mm but the AP1d was larger than 20 mm, except in one event (AP1d: 3.4 mm), which was characterised as the highest $IP_{max}30$ (i.e., 77.5 mm h^{-1}) of the study period. These event characteristics suggest that P_{tot} was the main factor controlling an effective hydrological response as antecedent conditions (i.e., soil moisture degree of the catchment) have been demonstrated to play an important role in R generation [18]. Similarly, Lana-Renault [10] identified a cluster of events characterized with the highest Q_0 that generated a larger R response, but the convective rainfall occurring in the Pyrenees during summer blurred the seasonal pattern described in the Es Fangar catchment.

Figure 7b shows that high Q_{max} values were observed in response to large rainfall amounts (above 50 mm) or for flood events with a Q_0 value higher than 0.080 m^3 s^{-1}. Both factors had a combined effect on the generation of high Q_{max} values as shown by the significant correlation observed between P_{tot} and Q_0 and Q_{max}. The role of wet conditions over Q_{max} was also observed in other Mediterranean-climate catchments where AP1d correlated significantly with Q_{max} [15,17,71].

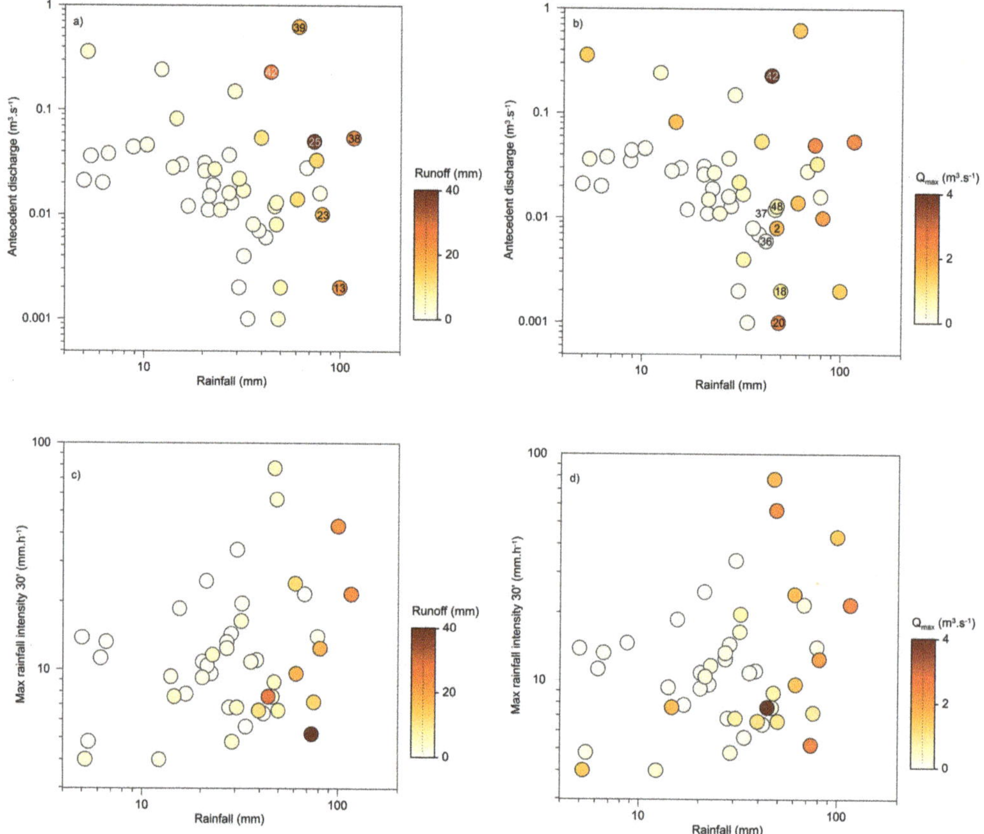

Figure 7. Multiples relationships between rainfall and R variables at the event scale at Es Fangar Creek: (**a**) rainfall-antecedent discharge-runoff; (**b**) rainfall–antecedent discharge-Q_{max} relationship; (**c**) $IP_{max}30$-rainfall-runoff and (**d**) $IP_{max}30$-rainfall-Q_{max}. The points with an ID are depicting the selected events for the non-linearity assessment, as it can be further consulted in Figure 6 and Table 3.

Table 2. Pearson correlation matrix between selected variables.

Variables	Q_0	P_{tot}	$IP_{mean}30$	$IP_{max}30$	Q_{max}	R	R_c
Q_0	1	0.03	0.00	0.08	**0.39**	0.28	**0.53**
P_{tot}		1	0.09	0.27	**0.55**	**0.68**	0.07
$IP_{mean}30$			1	**0.79**	0.15	−0.06	−0.21
$IP_{max}30$				1	0.25	0.03	−0.17
Q_{max}					1	**0.82**	**0.67**
R						1	**0.62**
R_c							1
Correlation significant at 0.01 level							

R and Q_{max} were not significantly correlated with $IP_{max}30$ (Table 2). Figure 7c,d show how high values of R and Q_{max} resulted for events with low $IP_{max}30$ when rainfall was > 50 mm or high Q_0. This absence of correlation between rainfall and the catchment response (R and Q_{max}) indicates that R generation is most likely the result of saturation excess processes [72], where P_{tot} and antecedent conditions played a key role. For events between June and September, saturation excess processes were, however, probably less dominant and combined with infiltration excess processes in response to highest $IP_{max}30$ values ranging from 19 to 56.4 mm h^{-1}. As a result, the events observed between June and September were shorter (5.7 to 14 h duration compared to 8 to 65.5 h for wet conditions events). Therefore, R mechanisms co-exist within a Mediterranean catchment [13] such as Es Fangar, with a marked seasonality, with saturation processes being dominant during the wet period and Hortonian R during the dry period. Consequently, the alternation of the R mechanisms generated lower and larger R contributions in events occurring during the dry and wet seasons, respectively. These processes caused the non-linearity of the hydrological response because the same amount of P_{tot} generated a non-lineal R response due to different soil moisture conditions [15,64].

4.2.3. Non-Linearity Assessment at Es Fangar Catchment

The events analysed in Figure 6 presented a huge variability in their hydrological response, despite recording a similar P_{tot}, ranging from 41.8 to 49.8 mm. R_c ranged from 1% to 65%, depending on the catchment moisture conditions, rainfall intensities and seasonality characteristics. The highest hydrological response occurred under marked wet soil moisture conditions in the winter period, even with low rainfall intensities (Figure 6a). A similar complexity in R responses has already been observed in other Mediterranean-climate catchments where seasonality and antecedent soil moisture conditions played a key role in R generation [11,57]. Gallart et al. [73] also showed that different R generation behaviours could be observed during the year due to varying catchment wetness conditions and changing rainfall events characteristics. The results obtained by these authors are clearly illustrated by comparing Events 2 and 20 with Events 18 and 37 in Figure 6. Indeed, R events that occurred during wet conditions with low rainfall intensities (18 and 37) showed similar R_c than events that occurred during dry conditions with high rainfall intensities (2 and 20). However, during dry conditions, runoff events resulting from low intensity rainfall yielded the lowest R responses (e.g., Event 36). Our results are also in agreement with the observations made by Schnabel et al. [74], who observed higher R contributions for low rainfall intensities events occurring in periods with high soil water content than for events with high rainfall intensities. Besides, those authors pointed out that this condition was common during years with above average rainfall (i.e., during wet years) but was rarely observed during dry years.

5. Conclusions

The evaluation of multiple temporal scales in contrasting small Mediterranean-climate catchments has improved the understanding of the role played by lithology and land use in the hydrological response. The assessment of rainfall-runoff relationships at the annual scale in small Mediterranean-climate catchments showed a significant linearity in the hydrological response due to the importance of the annual rainfall amount. Nevertheless, lithology effects on R generation explained an increase in the scattering in the rainfall-runoff relationship because pervious and impervious materials triggered larger and lower R contribution, respectively. Despite the significant correlation between rainfall and R, the Es Fangar Creek dataset illustrated a huge intra-annual variability of the rainfall-runoff relationship during the five hydrological years analysed, as seasonal rainfall and evapotranspiration dynamics controlled the R response. These dynamics were clearly observed in the average seasonal R_c, decreasing from winter to summer. As a result, these seasonal differences should be considered as a starting point of the non-linearity generation in the rainfall-runoff relationships at the event scale.

At the event scale, lineal and non-lineal relationships were observed in the rainfall-runoff relationships in small Mediterranean-climate catchments, suggesting that different factors conditioned the R response. The total rainfall was the most significant driving factor, although the interaction between seasonality and the spatial diversity of lithology and land uses at the catchment scale also played an important role in R generation. Thus, the highest correlations at the seasonal scale were observed in those events which occurred in winter and spring when the highest water reserves favoured the R response. Lithology caused a higher dispersion in the rainfall-runoff relationships at the event scale in the set of small Mediterranean catchments because pervious materials required higher antecedent wetness conditions. Different land uses promoted a decrease in R generation, comparing agricultural with scrubland uses, because agriculture promoted the highest correlation in the rainfall-runoff relationships due to lower vegetation cover. Therefore, human-induced alterations related to land uses in the R generation must be assessed in those countries were the abandonment of upland agricultural catchments have allowed natural vegetation to expand (i.e., southern Europe) and also in countries were agriculture and irrigated areas are increasingly replacing forest and shrub lands (i.e., North African Maghreb). However, events under agricultural land use or which occurred in the winter season independently of the land use and lithology catchments were the situations which were able to generate a higher linearity in the rainfall-runoff relationships.

This temporal downscaling from the annual to event scale elucidated how different R mechanisms can co-exist in small Mediterranean-climate catchments, considering the main temporal and spatial factors that govern the river catchment connectivity. Despite this, controls on R generation in Mediterranean-climate catchments require further attention to assess the role of lithology, land use and seasonality and their combined effects on the hydrological response for going beyond in the comprehension of highly sensitive areas to global change, such as the Mediterranean region. Considering that the performance gaps of hydrological processes in hydrological models due to the results of rainfall-runoff at daily scales are smoothed in small Mediterranean-climate catchments, the findings of this study improve the comprehension of hydrological processes as practical field knowledge needed to complete hydrological modelling at the event scale.

Author Contributions: Conceptualization, J.F., J.L. and J.E.; methodology J.E. and M.T.-B.; data curation, J.G.-C., A.C. and M.T.-B.; writing-original draft preparation J.F.; writing-review and editing, J.F., J.L., J.C. and J.E.; visualization J.G.-C., J.C. and A.C. and supervision J.L. and J.E. All authors have read and agreed to the published version of the manuscript.

Funding: This work was supported by the research project CGL2017-88200-R "Functional hydrological and sediment connectivity at Mediterranean catchments: global change scenarios –MEDhyCON2" funded by the Spanish Ministry of Science, Innovation and Universities, the Spanish Agency of Research (AEI) and the European Regional Development Funds (ERDF). The contribution of Jérôme Latron was supported by the research project PCIN-2017-061/AEI also funded by the Spanish Government. Josep Fortesa has a contract funded by the Ministry of Innovation, Research and Tourism of the Autonomous Government of the Balearic Islands (FPI/2048/2017). Julián García-Comendador is in receipt of a pre-doctoral contract (FPU15/05239) funded by the Spanish Ministry of Education, Culture and Sport. Miquel Tomàs-Burguera acknowledges the support from the project CGL2017-83866-C3-3-R financed by the European Regional Development Funds (ERDF) and the Spanish Ministry of Science, Innovation and Universities. Jaume Company is in receipt of Young Qualified Program fund by Employment Service of the Balearic Islands and European Social Fund (SJ-QSP 48/19). Aleix Calsamiglia acknowledges the support from the Spanish Ministry of Science, Innovation and Universities through a pre-doctoral contract (BES-2013-062887).

Acknowledgments: Part of the meteorological data were provided by the Spanish Meteorological Agency (AEMET).

Appendix A

Table A1. Mediterranean studies used into the rainfall-runoff relationship at (a) annual and (b) event scale.

Time Scale	Catchment	Country	Area (km²)	Precipitation (mm year⁻¹)	Temperature (°C)	Main Lithology	Reference
	Watershed C	USA	0.05	657	16.1	Impervious	Cited in Latron [47]
	TM9	Spain	0.06	891	13.5	Impervious	Cited in Latron [47]
	Watershed B	USA	0.09	901.7	13.9	Impervious	Dahlgren et al. [75]
	Watershed G	USA	0.13	931.3	13.9	Impervious	Dahlgren et al. [75]
	Watershed B	USA	0.17	623	16.1	Impervious	Cited in Latron [47]
	Watershed A	USA	0.19	623	16.1	Impervious	Cited in Latron [47]
	Spruce	France	0.20	1794	6.6	Impervious	Didon-Lescot et al. [36]
	La Sapine	France	0.20	1715.6	6.6	Impervious	Didon-Lescot et al. [36]
	Watershed C	USA	0.23	898.8	13.9	Impervious	Dahlgren et al. [75]
	Mokobulaam	South Africa	0.26	1150	17.2	Impervious	Cited in Nadal-Romero [34]
	HREC	USA	0.33	549	15.0	Pervious	Swarowsky et al. [54]
	Guadalperalón	Spain	0.35	502	16.0	Pervious	Ceballos and Schnabel [9]
	La Teula	Spain	0.39	595.5	12.5	Impervious	Cited in Latron [47]
	Araguás	Spain	0.45	764.6	11.8	Impervious	Nadal-Romero [34]
	Avic	Spain	0.52	548	12.5	Impervious	Cited in Latron [47]
(a) Annual scale	Sta Magdalena	Spain	0.53	1418	7.3	Pervious	Latron [47]
	Can Vila	Spain	0.56	1041	9.0	Impervious	Latron [47]
	Boussicaut	France	0.73	1038	13.9	Impervious	Cited in Latron [47]
	San Salvador	Spain	0.92	1108	9.5	Pervious	Serrano-Muela [56]
	Parapuños	Spain	1.00	516.2	14.3	Impervious	Schnabel et al. [74]
	Schubert	USA	1.03	708	15.7	Impervious	Lewis et al. [35]
	Can Revull	Spain	1.03	517	16.5	Impervious	Estrany et al. [15]
	Sa Murtera	Spain	1.20	520.4	16.5	Pervious	García-Comendador et al. [58]
	Cannata	Italy	1.30	671	14.3	Pervious	Licciardello et al. [76]
	Ca l'Isard	Spain	1.32	1128	7.3	Pervious	Latron [47]
	Vaubarnier	France	1.49	1059	13.9	Impervious	Cited in Latron [47]
	Desteou	France	1.53	1169	13.9	Impervious	Cited in Latron [47]
	Bosc	Spain	1.60	675.1	13.0	Impervious	Pacheco et al. [77]
	TM0	Spain	2.00	1004	13.5	Impervious	Cited in Latron [47]
	Campàs	Spain	2.57	675.1	13.0	Impervious	Pacheco et al. [77]
	Kamech	Tunisia	2.63	650	14.0	Impervious	Slimane et al. [78]

Table A1. *Cont.*

Time Scale	Catchment	Country	Area (km^2)	Precipitation (mm year^{-1})	Temperature (°C)	Main Lithology	Reference
	Arnás	Spain	2.84	951	11.0	Impervious	Lana-Renalut [10]
	Headwater 4	Jordan	3.20	493	16.7	Pervious	Riest et al. [61]
	Es Fangar	Spain	3.35	840.8	15.7	Pervious	This study
	Cal Rodó	Spain	4.17	1125	9.0	Pervious	Latron [47]
	Headwater 3	Jordan	4.20	491	16.7	Pervious	Riest et al. [61]
	Biniaraix	Spain	4.40	1453.2	12.0	Pervious	Calsamiglia et al. [55]
	Sa Font de la Vila	Spain	4.80	519.6	16.5	Pervious	García-Comendador et al. [58]
	Cogolins	France	5.47	833	13.9	Impervious	Cited in Latron [47]
	Headwater 2	Jordan	8.40	367	16.7	Pervious	Riest et al. [61]
	Maurets	France	8.48	1088	13.9	Impervious	Cited in Latron [47]
	Valescure	France	9.22	1208	13.9	Impervious	Cited in Latron [47]
	Maraval	France	9.61	822	13.9	Impervious	Cited in Latron [47]

	Catchment	Country	Area (km^2)	Precipitation (mm year^{-1})	Land use	Main lithology	Reference
	Guadalperalón	Spain	0.35	22	Agroforestry	Pervious	Ceballos and Schnabel [9]
	Can Vila	Spain	0.56	39.6	Agricultural	Impervious	Latron and Gallart [18]; Roig-Planasdemunt et al. [79]; Cayuelta et al. [80]
(b) Event scale	Parapuños	Spain	1.00	17.4	Agroforestry	Impervious	Schnabel et al. [74]
	Can Revull	Spain	1.03	19.9	Agricultural	Impervious	Estrany et al. [15]
	Sa Murtera	Spain	1.20	43.9	Shrub	Pervious	García-Comendador et al. [58]
	Arnás	Spain	2.84	20.1	Shrub	Impervious	Lana-Renalut [10]
	Headwater 4	Jordan	3.20	225	Agricultural	Pervious	Riest et al. [61]
	Es Fangar	Spain	3.35	29	Agroforestry	Pervious	This study
	Headwater 3	Jordan	4.20	213	Shrub	Pervious	Riest et al. [61]
	Biniaraix	Spain	4.40	42.6	Forest	Pervious	Calsamiglia et al. [55]
	Sa Font de la Vila	Spain	4.80	36.9	Shrub	Pervious	García-Comendador et al. [58]
	Headwater 2	Jordan	8.40	174	Shrub	Pervious	Riest et al. [61]

Table 2. Relative rainfall and R contribution of the highest 5 days and number of days to reach the 50% of R for each hydrological year in the Es Fangar Creek.

Year	Rainfall Contribution of 5 Days (%)	Runoff Contribution of 5 Days (%)	Number of Days to Reach 50% of Runoff
2012–2013	32	37	10
2013–2014	35	53	5
2014–2015	23	50	5
2015–2016	35	28	17
2016–2017	41	28	3

Table 3. Flood event characteristics during the study period 2012–2017 in the Es Fangar Creek.

ID	Date	P_{tot} (mm)	$IP_{mean}30$ (mm h^{-1})	$IP_{max}30$ (mm h^{-1})	Q_{dur} (h)	Q_0 (m^3 s^{-1})	Q_{max} (m^3 s^{-1})	R (mm)	R_c (%)	AP1d (mm)	AP3d (mm)
1	27 October 2012 20:15	30.6	10.1	33.8	8.0	0.002	0.144	0.3	1.0	0.0	0.0
2	11 November 2012 22:20	47.3	11.3	77.5	18.0	0.008	1.747	6.7	14.1	3.4	3.4
3	28 November 2012 08:00	15.5	6.2	18.6	17.5	0.030	0.094	1.0	6.7	9.7	9.7
4	24 January 2013 09:00	21.3	5.7	24.6	12.7	0.011	0.036	0.3	1.6	0.0	15.3
5	24 January 2013 22:30	5.0	6.7	13.8	15.7	0.021	0.083	0.7	13.9	21.3	36.6
6	28 January 2013 03:30	16.8	4.5	7.8	44.5	0.012	0.064	1.2	6.9	1.7	32.5
7	28 February 2013 08:00	32.1	3.9	16.4	21.2	0.017	0.427	4.5	14.0	4.6	11.0
8	01 March 2013 21:00	8.7	5.0	14.6	8.0	0.035	0.077	0.0	0.5	32.2	40.6
9	13 March 2013 14:15	60.6	8.7	24.0	12.8	0.014	1.610	12.8	21.1	0.0	0.0
10	14 March 2013 20:30	6.6	8.8	13.2	17.6	0.038	0.156	1.6	23.5	60.8	60.8
11	05 April 2013 22:30	38.5	4.1	11.0	48.5	0.007	0.059	1.4	3.6	0.0	0.0
12	28 April 2013 21:00	24.6	ND	ND	46.5	0.011	0.347	3.0	12.3	22.1	23.0
13	19 November 2013 02:15	99.3	5.3	43.0	64.7	0.002	1.414	22.5	22.7	7.0	75.5
14	22 November 2013 09:30	8.8	ND	ND	24.7	0.044	0.093	1.6	18.2	1.3	1.3
15	27 November 2013 18:45	14.0	3.1	9.3	39.5	0.028	0.115	2.6	18.6	9.2	15.3
16	29 November 2013 10:30	10.3	ND	ND	25.5	0.046	0.155	2.4	23.3	3.6	28.8
17	30 November 2013 18:45	39.5	2.3	6.6	46.3	0.054	1.105	11.3	28.7	10.3	40.2
18	04 April 2014 03:15	49.8	2.6	6.6	44.7	0.002	1.006	7.8	15.6	3.8	8.0
19	25 April 2014 06:45	28.6	6.4	14.4	15.5	0.000	0.180	0.8	2.9	5.0	13.4
20	29 September 2014 13:45	48.6	16.2	56.4	14.0	0.001	2.356	4.2	8.7	11.8	13.9
21	04 December 2014 22:45	27.2	4.5	12.4	44.3	0.016	0.197	2.1	7.9	9.4	69.2
22	28 December 2014 10:45	36.0	5.1	10.8	39.3	0.008	0.183	2.5	7.0	2.2	2.2
23	20 January 2015 13:00	81.0	2.7	12.4	50.0	0.010	2.094	17.3	21.4	9.0	9.0
24	03 February 2015 09:45	22.6	1.4	9.6	7.5	0.019	0.058	0.3	1.5	0.4	21.6
25	04 February 2015 11:15	73.6	1.3	5.2	65.5	0.050	2.453	39.2	53.2	22.6	35.8
26	21 February 2015 22:15	30.6	1.6	6.8	24.8	0.022	0.883	5.9	19.3	0.0	0.0
27	24 February 2015 00:45	20.4	2.7	10.8	18.2	0.031	0.077	1.0	4.7	0.0	30.6
28	27 February 2015 13:00	20.4	3.1	9.2	31.8	0.026	0.164	2.1	10.3	5.4	11.8
29	14 March 2015 18:00	21.6	3.9	10.4	16.7	0.015	0.351	1.7	8.1	0.0	0.0
30	22 March 2015 02:45	6.2	4.1	11.2	8.7	0.020	0.056	0.3	5.3	1.4	11.6
31	24 March 2015 23:15	75.6	1.8	7.2	42.8	0.033	0.924	13.9	18.4	0.0	7.6
32	04 September 2015 03:00	32.4	1.7	19.6	11.0	0.004	0.783	1.1	3.3	0.6	2.6
33	30 September 2015 08:15	34.0	2.4	5.6	9.7	0.001	0.132	0.7	2.0	8.0	13.0
34	30 September 2015 18:15	27.2	2.0	13.2	16.8	0.037	0.207	1.3	4.9	36.2	47.0
35	01 April 2016 16:00	78.8	2.7	14.0	43.5	0.016	0.222	2.4	3.0	0.0	0.0
36	07 May 2016 11:15	41.8	2.7	6.4	13.5	0.006	0.092	0.5	1.2	1.6	1.6
37	19 December 2016 02:30	46.4	2.3	7.6	27.2	0.012	0.394	4.4	9.5	2.0	16.6
38	20 December 2016 09:00	115.8	6.3	21.6	16.0	0.054	2.434	23.5	20.3	24.4	49.2
39	21 December 2016 04:30	61.4	4.9	9.6	33.7	0.626	1.565	17.6	28.6	116.0	164.6
40	20 January 2017 07:50	28.0	1.8	6.8	13.0	0.013	0.180	1.4	5.0	4.0	9.8
41	20 January 2017 17:45	14.6	1.7	7.6	10.7	0.082	1.687	5.9	40.2	28.4	37.2
42	21 January 2017 07:00	44.2	2.6	7.6	19.2	0.230	3.942	28.9	65.4	35.4	50.6
43	22 January 2017 04:00	5.2	1.0	4.0	5.7	0.360	1.465	4.2	80.8	35.4	50.6
44	22 January 2017 16:00	12.2	0.8	4.0	9.0	0.243	0.461	2.9	23.8	44.4	96.2
45	23 January 2017 20:52	29.0	2.1	4.8	25.0	0.150	0.355	4.8	16.5	20.2	87.2
46	25 January 2017 11:45	5.4	1.2	4.8	31.3	0.036	0.235	0.3	5.4	0.0	51.6
47	27 January 2017 17:00	23.0	2.9	11.6	55.0	0.027	0.306	5.6	24.5	0.0	5.6
48	8 February 2017 06:30	47.4	2.9	8.8	41.5	0.013	0.843	5.6	11.8	0.0	12.8
49	5 June 2017 10:45	67.6	3.0	21.6	5.7	0.028	0.330	0.4	0.7	2.6	2.6
	Min	5.0	0.8	4.0	5.7	0.000	0.036	0.0	0.5	0.0	0.0
	Max	115.8	16.2	77.5	65.5	0.626	3.942	39.2	80.8	116.0	164.6
	Median	29.0	2.9	10.8	19.2	0.0	0.3	2.4	11.8	4.0	13.9
	SD	25.5	3.0	13.8	16.4	0.108	0.857	8.3	16.2	20.6	31.6

Table 4. Main characteristics of the selected events for non-linearity analysis in Es Fangar Creek.

ID	Flood Event	Ptot (mm)	$IP_{max}30$ (mm h^{-1})	R (mm)	R_c (%)	Q_{max} (m^3 s^{-1})	Q_0 (m^3 s^{-1})	AP1d (mm)
42	21 January 2017 07:00	44.2	7.6	28.9	65	3.942	0.229	35.4
18	04 April 2014 03:15	49.8	6.6	7.8	16	1.006	0.002	3.8
2	11 November 2012 22:15	47.3	77.5	6.7	14	1.747	0.008	3.4
48	08 February 2017 06:30	47.4	8.8	5.6	12	0.843	0.012	0.0
37	19 December 2016 02:30	46.4	7.6	4.4	9	0.394	0.012	2.0
20	29 September 2014 13:45	48.6	56.4	4.2	9	2.356	0.015	11.8
36	07 May 2016 11:15	41.8	6.4	0.5	1	0.092	0.006	1.6

Table 5. Main characteristics of the highest runoff event contribution in Es Fangar Creek during the study period 2012–2017.

ID	Flood Event	Ptot (mm)	$IP_{max}30$ (mm h^{-1})	R (mm)	R_c (%)	Q_{max} (m^3 s^{-1})	Q_0 (m^3 s^{-1})	AP1d (mm)
25	04 February 2015 11:15	73.6	5.2	39.2	53	2.453	0.046	22.6
42	21 January 2017 07:00	44.2	7.6	28.9	65	3.942	0.229	35.4
38	20 December 2016 09:00	115.8	21.6	23.5	20	2.434	0.054	24.4
13	19 November 2013 02:15	99.3	43.0	22.5	23	1.414	0.021	7.0
39	21 December 2016 04:30	61.4	9.6	17.6	29	1.565	0.626	116.0
23	20 January 2015 13:00	81.0	12.4	17.3	21	2.094	0.010	9.0

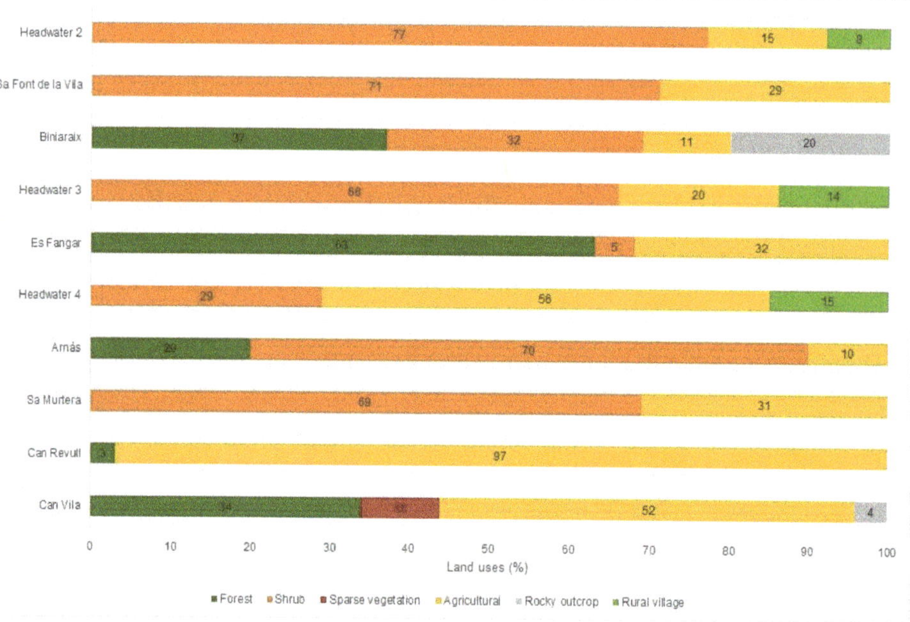

Figure 1. Land uses (%) for the 10 of 12 small Mediterranean-climate catchments selected for the rainfall-runoff relationship analysis. Guadalperalón and Parapuños catchments have not detailed land uses information.

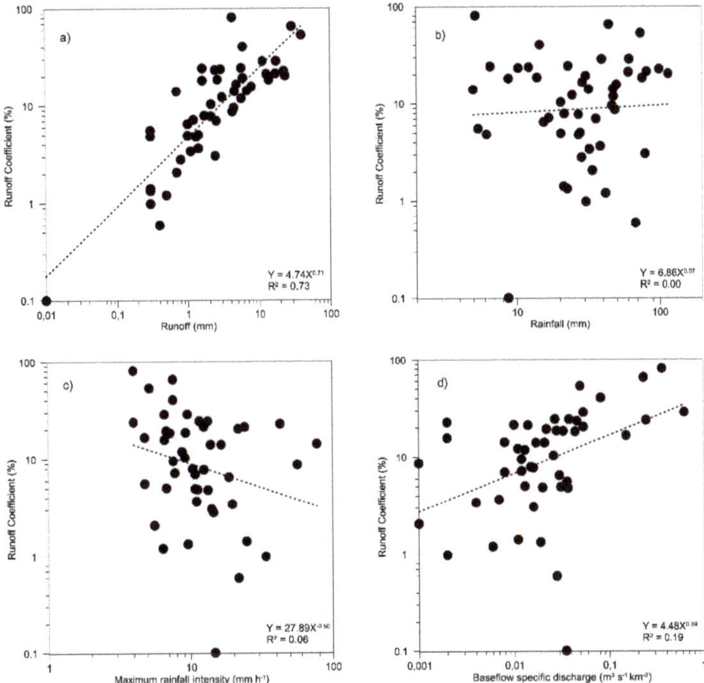

Figure 2. Relationship between (**a**) R and R_c coefficient, (**b**) rainfall and R_c, (**c**) $IP_{max}30$ intensity and R_c and (**d**) base-flow specific discharge and R_c at Es Fangar Creek. Dotted lines show significant ($p < 0.01$) fits with a power function.

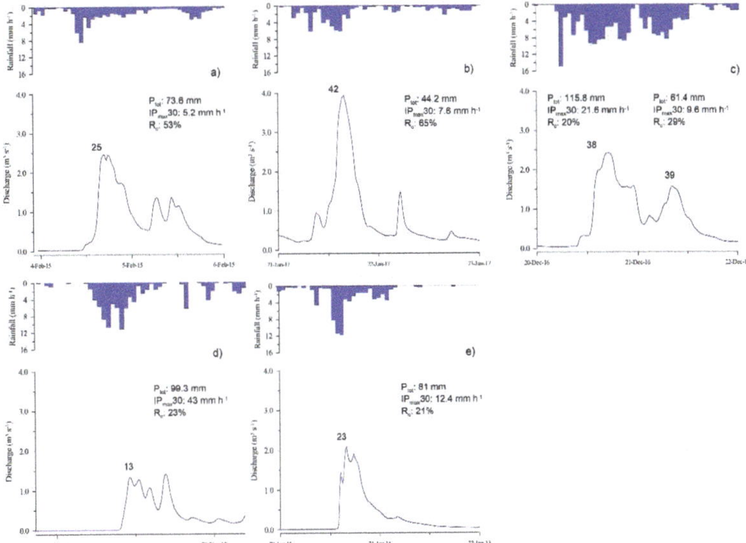

Figure 3. Highest R event contribution during the study period (2012–17) at Es Fangar Creek. The numbers located at the peak of each hydrograph indicate the ID of the events recorded in Es Fangar Creek during the study period 2012–2017, further exposed in Table 3.

References

1. Chiu, M.-C.; Leigh, C.; Mazor, R.; Cid, N.; Resh, V. Anthropogenic Threats to Intermittent Rivers and Ephemeral Streams. In *Intermittent Rivers and Ephemeral Streams: Ecology and Management*; Academic Press: Cambridge, MA, USA, 2017; pp. 433–454, ISBN 978-0-12-803904-5.
2. Borg Galea, A.; Sadler, J.P.; Hannah, D.M.; Datry, T.; Dugdale, S.J. Mediterranean Intermittent Rivers and Ephemeral Streams: Challenges in monitoring complexity. *Ecohydrology* **2019**, *12*. [CrossRef]
3. Kottek, M.; Grieser, J.; Beck, C.; Rudolf, B.; Rubel, F. World Map of the Köppen-Geiger climate classification updated. *Meteorol. Z.* **2006**, *15*, 259–263. [CrossRef]
4. Datry, T.; Bonada, N.; Boulton, A. General Introduction. In *Intermittent Rivers and Ephemeral Streams: Ecology and Management*; Academic Press: Cambridge, MA, USA, 2017; pp. 1–597. ISBN 978-0-12-803904-5. Available online: https://www.elsevier.com/books/intermittent-rivers-and-ephemeral-streams/datry/978-0-12-803835-2 (accessed on 18 January 2020).
5. Oueslati, O.; De Girolamo, A.M.; Abouabdillah, A.; Kjeldsen, T.R.; Lo Porto, A. Classifying the flow regimes of Mediterranean streams using multivariate analysis. *Hydrol. Process.* **2015**, *29*, 4666–4682. [CrossRef]
6. Costigan, K.H.; Kennard, M.J.; Leigh, C.; Sauquet, E.; Boulton, A.J. Flow Regimes in Intermittent Rivers and Ephemeral Streams. *Intermittent Rivers Ephemer Streams.* 2017, pp. 51–78. Available online: https://www.elsevier.com/books/intermittent-rivers-and-ephemeral-streams/datry/978-0-12-803835-2 (accessed on 18 January 2020).
7. De Girolamo, A.M.; Lo Porto, A.; Pappagallo, G.; Tzoraki, O.; Gallart, F. The Hydrological Status Concept: Application at a Temporary River (Candelaro, Italy). *River Res. Appl.* **2015**, *31*, 892–903. [CrossRef]
8. Latron, J.; Soler, M.; Llorens, P.; Gallart, F. Spatial and temporal variability of the hydrological response in a small Mediterranean research catchment (Vallcebre, Eastern Pyrenees). *Hydrol. Process.* **2008**, *22*, 775–787. [CrossRef]
9. Ceballos, A.; Schnabel, S. Hydrological behaviour of a small catchment in the dehesa landuse system (Extremadura, SW Spain). *J. Hydrol.* **1998**, *210*, 146–160. [CrossRef]
10. Lana-Renault, N. Respuesta Hidrológica y Sedimentológica en una Cuenca de Montaña Media Afectada por Cambios de Cubierta Vegetal: La Cuenca Experimental de Arnás. Pirineo Central, Universidad de Zaragoza. 2007. Available online: https://dialnet.unirioja.es/servlet/tesis?codigo=113822 (accessed on 21 April 2019).
11. López-Tarazón, J.; Batalla, R.J.; Vericat, D.; Balasch, J.C. Rainfall, runoff and sediment transport relations in a mesoscale mountainous catchment: The River Isábena (Ebro basin). *Catena* **2010**, *82*, 23–34. [CrossRef]
12. Lana-Renault, N.; Latron, J.; Regüés, D. Streamflow response and water-table dynamics in a sub-Mediterranean research catchment (Central Pyrenees). *J. Hydrol.* **2007**, *347*, 497–507. [CrossRef]
13. Manus, C.; Anquetin, S.; Braud, I.; Vandervaere, J.-P.; Creutin, J.-D.; Viallet, P.; Gaume, E. Hydrology and Earth System Sciences A modeling approach to assess the hydrological response of small mediterranean catchments to the variability of soil characteristics in a context of extreme events. *Hydrol. Earth Syst. Sci.* **2009**, *13*, 79–97. [CrossRef]
14. Efstratiadis, A.; Koussis, A.D.; Koutsoyiannis, D.; Mamassis, N. Flood design recipes vs. reality: Can predictions for ungauged basins be trusted? *Hazards Earth Syst. Sci* **2014**, *14*, 1417–1428. [CrossRef]
15. Estrany, J.; Garcia, C.; Batalla, R.J. Hydrological response of a small mediterranean agricultural catchment. *J. Hydrol.* **2010**, *380*, 180–190. [CrossRef]
16. Huza, J.; Teuling, A.J.; Braud, I.; Grazioli, J.; Melsen, L.A.; Nord, G.; Raupach, T.H.; Uijlenhoet, R. Precipitation, soil moisture and runoff variability in a small river catchment (Ardèche, France) during HyMeX Special Observation Period 1. *J. Hydrol.* **2014**, *516*, 330–342. [CrossRef]
17. Zoccatelli, D.; Marra, F.; Armon, M.; Rinat, Y.; Smith, J.A.; Morin, E. Contrasting rainfall-runoff characteristics of floods in desert and Mediterranean basins. *Hydrol. Earth Syst. Sci.* **2019**, *23*, 2665–2678. [CrossRef]
18. Latron, J.; Gallart, F. Seasonal dynamics of runoff-contributing areas in a small mediterranean research catchment (Vallcebre, Eastern Pyrenees). *J. Hydrol.* **2007**, *335*, 194–206. [CrossRef]
19. Lexartza-Artza, I.; Wainwright, J. Hydrological connectivity: Linking concepts with practical implications. *Catena* **2009**, *79*, 146–152. [CrossRef]
20. McGlynn, B.L.; McDonnell, J.J. Role of discrete landscape units in controlling catchment dissolved organic carbon dynamics. *Water Resour. Res.* **2003**, *39*, 1090. [CrossRef]
21. Buttle, J.M.; McDonald, D.J. Coupled vertical and lateral preferential flow on a forested slope. *Water Resour. Res.* **2002**, *38*, 18-1–18-16. [CrossRef]

22. Tetzlaff, D.; Soulsby, C.; Waldron, S.; Malcolm, I.A.; Bacon, P.J.; Dunn, S.M.; Lilly, A.; Youngson, A.F. Conceptualization of runoff processes using a geographical information system and tracers in a nested mesoscale catchment. *Hydrol. Process.* **2007**, *21*, 1289–1307. [CrossRef]

23. Emanuel, R.E.; Hazen, A.G.; Mcglynn, B.L.; Jencso, K.G. Vegetation and topographic influences on the connectivity of shallow groundwater between hillslopes and streams. *Ecohydrology* **2014**, *7*, 887–895. [CrossRef]

24. Boulton, A.J.; Rolls, R.J.; Jaeger, K.L. Hydrological Connectivity in Intermittent Rivers and Ephemeral Streams. In *Intermittent Rivers Ephemer Streams*; Academic Press: Cambridge, MA, USA, 2017; pp. 79–108. Available online: https://www.sciencedirect.com/science/article/pii/B9780128038352000048 (accessed on 18 January 2020).

25. Penna, D.; Tromp-Van Meerveld, H.J.; Gobbi, A.; Borga, M.; Dalla Fontana, G. The influence of soil moisture on threshold runoff generation processes in an alpine headwater catchment. *Hydrol. Earth Syst. Sci.* **2011**, *15*, 689–702. [CrossRef]

26. Tromp-Van Meerveld, H.J.; McDonnell, J.J. Threshold relations in subsurface stormflow: 1. A 147-storm analysis of the Panola hillslope. *Water Resour. Res.* **2006**, *42*, W02410. [CrossRef]

27. Saffarpour, S.; Western, A.W.; Adams, R.; Mcdonnell, J.J. Multiple runoff processes and multiple thresholds control agricultural runoff generation. *Hydrol. Earth Syst. Sci* **2016**, *20*, 4525–4545. [CrossRef]

28. Woodward, J. (Ed.) *The Physical Geography of the Mediterranean*; Oxford University Press: New York, NY, USA, 2009; ISBN 978-0-19-926803-0.

29. Coustau, M.; Bouvier, C.; Borrell-Estupina, V.; Jourde, H. Natural Hazards and Earth System Sciences Flood modelling with a distributed event-based parsimonious rainfall-runoff model: Case of the karstic Lez river catchment. *Hazards Earth Syst. Sci* **2012**, *12*, 1119–1133. [CrossRef]

30. Koutroulis, A.G.; Tsanis, I.K.; Daliakopoulos, I.N.; Jacob, D. Impact of climate change on water resources status: A case study for Crete Island, Greece. *J. Hydrol.* **2013**, *479*, 146–158. [CrossRef]

31. Nikolaidis, N.P.; Demetropoulou, L.; Froebrich, J.; Jacobs, C.; Gallart, F.; Prat, N.; Porto, A.L.; Campana, C.; Papadoulakis, V.; Skoulikidis, N.; et al. Towards sustainable management of Mediterranean river basins: Policy recommendations on management aspects of temporary streams. *Water Policy* **2013**, *15*, 830–849. [CrossRef]

32. Calvo-Cases, A.; Boix-Fayos, C.; Imeson, A. Runoff generation, sediment movement and soil water behaviour on calcareous (limestone) slopes of some Mediterranean environments in southeast Spain. *Geomorphology* **2003**, *50*, 269–291. [CrossRef]

33. Cantón, Y.; Solé-Benet, A.; De Vente, J.; Boix-Fayos, C.; Calvo-Cases, A.; Asensio, C.; Puigdefábregas, J. A review of runoff generation and soil erosion across scales in semiarid south-eastern Spain. *J. Arid Environ.* **2011**, *75*, 1254–1261. [CrossRef]

34. Nadal-Romero, E. Las áreas de Cárcavas (badlands) Como Fuente de Sedimento en Cuencas de Montaña: Procesos de Meteorización, Erosión y Transporte en Margas del Pirineo Central, Universidad de Zaragoza. 2007. Available online: https://dialnet.unirioja.es/servlet/tesis?codigo=206095 (accessed on 22 April 2019).

35. Lewis, D.; Singer, M.; Dahlgren, R.; Tate, K. Hydrology in a California oak woodland watershed: A 17-year study. *J. Hydrol.* **2000**, *240*, 106–117. [CrossRef]

36. Didon-Lescot, J.-F.; Guillet, B.; Lelong, F. Effect of the clearfelling on the water quality: Example of a spruce forest on a small catchment in France. *Acta Geológica Hispánica* **1993**, *28*, 45–53.

37. Zuazo, D.V.; Pleguezuelo, C. Soil-erosion and runoff prevention by plant covers. A review. *Agron. Sustain. Dev.* **2008**, *28*, 65–86. [CrossRef]

38. Buendia, C.; Batalla, R.J.; Sabater, S.; Palau, A.; Marcé, R. Runoff Trends Driven by Climate and Afforestation in a Pyrenean Basin. *Land Degrad. Dev.* **2016**, *27*, 823–838. [CrossRef]

39. Caloiero, T.; Biondo, C.; Callegari, G.; Collalti, A.; Froio, R.; Maesano, M.; Matteucci, G.; Pellicone, G.; Veltri, A. Results of a long-term study on an experimental watershed in southern Italy. *Forum Geogr.* **2016**, *15*, 55–65. [CrossRef]

40. Nadal-Romero, E.; Cammeraat, E.; Serrano-Muela, M.P.; Lana-Renault, N.; Regüés, D. Hydrological response of an afforested catchment in a Mediterranean humid mountain area: A comparative study with a natural forest. *Hydrol. Process.* **2016**, *30*, 2717–2733. [CrossRef]

41. Tarolli, P.; Preti, F.; Romano, N. Terraced landscapes: From an old best practice to a potential hazard for soil degradation due to land abandonment. *Anthropocene* **2014**, *6*, 10–25. [CrossRef]

42. Lesschen, J.P.; Cammeraat, L.H.; Nieman, T. Erosion and terrace failure due to agricultural land abandonment in a semi-arid environment. *Earth Surf. Process. Landforms* **2008**, *33*, 1574–1584. [CrossRef]

43. Arnáez, J.; Lana-Renault, N.; Lasanta, T.; Ruiz-Flaño, P.; Castroviejo, J. Effects of farming terraces on hydrological and geomorphological processes. A review. *Catena* **2015**, *128*, 122–134. [CrossRef]

44. Beven, K. *Rainfall-Runoff Modelling The Primer*, 2nd ed.; Wiley: Chichester, UK, 2012; ISBN 978-0-47-071459-1.

45. Klemeš, V. Dilettantism in hydrology: Transition or destiny? *Water Resour. Res.* **1986**, *22*, 177S–188S. [CrossRef]

46. Merheb, M.; Moussa, R.; Abdallah, C.; Colin, F.; Perrin, C.; Baghdadi, N. Hydrological response characteristics of Mediterranean catchments at different time scales: A meta-analysis. *Hydrol. Sci. J.* **2016**, *61*, 2520–2539. [CrossRef]

47. Latron, J. Estudio del Funcionamiento Hidrológico de una Cuenca Mediterránea de Montaña (Vallcebre, Pirineos Catalanes), Universitat de Barcelona. 2003. Available online: https://dialnet.unirioja.es/servlet/tesis?codigo=235272 (accessed on 18 April 2019).

48. Fick, S.E.; Hijmans, R.J. WorldClim 2: New 1-km spatial resolution climate surfaces for global land areas. *Int. J. Climatol.* **2017**, *37*, 4302–4315. [CrossRef]

49. Guijarro, J.A. Contribución a la Bioclimatología de Baleares, Universitat de les Illes Balears. 1986. Available online: https://dialnet.unirioja.es/servlet/libro?codigo=613782 (accessed on 14 March 2019).

50. YACU Estudio de Caracterización del régimen Extremo de Precipitaciones en la Isla de Mallorca, Junta D'Aigües de Les Illes Balears: Palma de Mallorca. 2003. Available online: http://observatoriaigua.uib.es/repositori/tp_precipitacion_2002.pdf (accessed on 18 April 2019).

51. Hargreaves, G.H.; Samani, Z.A. Reference Crop Evapotranspiration from Temperature. *Appl. Eng. Agric.* **1985**, *1*, 96–99. [CrossRef]

52. Maidment, D.R. *Handbook of Hydrology*, 1st ed.; McGraw-Hill: New York NY, USA, 1993; ISBN 0070397325. Available online: https://www.abebooks.com/9780070397323/Handbook-Hydrology-Maidment-David-0070397325/plp (accessed on 16 March 2019).

53. Gallart, F.; Amaxidis, Y.; Botti, P.; Canè, G.; Castillo, V.; Chapman, P.; Froebrich, J.; García-Pintado, J.; Latron, J.; Llorens, R.; et al. Investigating hydrological regimes and processes in a set of catchments with temporary waters in Mediterranean Europe. *Hydrol. Sci. J.* **2008**, *53*, 618–628. [CrossRef]

54. Swarowsky, A.; Dahlgren, R.A.; Tate, K.W.; Hopmans, J.W.; O'Geen, A.T. Catchment-Scale Soil Water Dynamics in a Mediterranean-Type Oak Woodland. *Vadose Zone J.* **2011**, *10*, 800. [CrossRef]

55. Calsamiglia, A.; Fortesa, J.; Garcia-Comendador, J.; Estrany, J. Respuesta hidro-sedimentaria en dos cuencas mediterráneas representativas afectadas por el cambio global. *Cuaternario y Geomorfol.* **2016**, *30*, 87–103. [CrossRef]

56. Serrano-Muela, M.P. Influencia de la Cubierta Vegetal y las Propiedades del Suelo en la Respuesta Hidrológica: Generación de Escorrentía en una Cuenca Forestal de la Montaña Media Pirenaica, Universidad de Zaragoza. 2012. Available online: https://dialnet.unirioja.es/servlet/tesis?codigo=77003 (accessed on 15 May 2019).

57. Gallart, F.; Llorens, P.; Latron, J.; Regüés, D. Hydrological processes and their seasonal controls in a small Mediterranean mountain catchment in the Pyrenees. *Hydrol. Earth Syst. Sci.* **2002**, *6*, 527–537. [CrossRef]

58. García-Comendador, J.; Fortesa, J.; Calsamiglia, A.; Calvo-Cases, A.; Estrany, J. Post-fire hydrological response and suspended sediment transport of a terraced Mediterranean catchment. *Earth Surf. Process. Landforms* **2017**, *42*, 2254–2265. [CrossRef]

59. Gaillard, E.; Lavabre, J.; Isbérie, C.; Normand, M. Etat hydrique d'une parcelle et écoulements dans un petit basin versant du massif cristallin des Maures. *Hydrogéologie* **1995**, *4*, 41–48.

60. Tzoraki, O.; Nikolaidis, N.P. A generalized framework for modeling the hydrologic and biogeochemical response of a Mediterranean temporary river basin. *J. Hydrol.* **2007**, *346*, 112–121. [CrossRef]

61. Ries, F.; Schmidt, S.; Sauter, M.; Lange, J. Controls on runoff generation along a steep climatic gradient in the Eastern Mediterranean. *J. Hydrol. Reg. Stud.* **2017**, *9*, 18–33. [CrossRef]

62. Regüés, D.; Gallart, F. Seasonal patterns of runoff and erosion responses to simulated rainfall in badland area in Mediterranean mountain conditions (Vallcebre, southeastern Pyrenees). *Earth Surf. Process. Landforms* **2004**, *29*, 755–767. [CrossRef]

63. Latron, J.; Gallart, F. Hydrological Response of Two Nested Small Mediterranean Basins Presenting Various Degradation States. *Phys. Chem. Earth* **1995**, *20*, 369–374. [CrossRef]

64. Nadal-Romero, E.; Latron, J.; Lana-Renault, N.; Serrano-Muela, P.; Martí-Bono, C.; Regüés, D. Temporal variability in hydrological response within a small catchment with badland areas, central Pyrenees. *Hydrol. Sci. J.* **2008**, *53*, 629–639. [CrossRef]

65. Cerdan, O.; Le Bissonnais, Y.; Govers, G.; Lecomte, V.; van Oost, K.; Couturier, A.; King, C.; Dubreuil, N. Scale effect on runoff from experimental plots to catchments in agricultural areas in Normandy. *J. Hydrol.* **2004**, *299*, 4–14. [CrossRef]

66. García-Ruiz, J.M.; Regüés, D.; Alvera, B.; Lana-Renault, N.; Serrano-Muela, P.; Nadal-Romero, E.; Navas, A.; Latron, J.; Martí-Bono, C.; Arnáez, J. Flood generation and sediment transport in experimental catchments affected by land use changes in the central Pyrenees. *J. Hydrol.* **2008**, *356*, 245–260. [CrossRef]

67. Beguería, S.; López-Moreno, J.I.; Lorente, A.; Seeger, M.; García-Ruiz, J.M. Assessing the Effect of Climate Oscillations and Land-use Changes on Streamflow in the Central Spanish Pyrenees. *AMBIO A J. Hum. Environ.* **2003**, *32*, 283–286. [CrossRef] [PubMed]

68. Buendia, C.; Bussi, G.; Tuset, J.; Vericat, D.; Sabater, S.; Palau, A.; Batalla, R.J. Effects of afforestation on runoff and sediment load in an upland Mediterranean catchment. *Sci. Total Environ.* **2016**, *540*, 144–157. [CrossRef]

69. Gallart, F.; Llorens, P. Water resources and environmental change in Spain. A key issue for sustainable integrated catchment management. *Cuad. Investig. Geográfica* **2001**, *27*, 7–16. [CrossRef]

70. Jordan, J.P. Spatial and temporal variability of stormflow generation processes on a Swiss catchment. *J. Hydrol.* **1994**, *1–4*, 357–382. [CrossRef]

71. Tuset, J.; Vericat, D.; Batalla, R.J. Rainfall, runoff and sediment transport in a Mediterranean mountainous catchment. *Sci. Total Environ.* **2016**, *540*, 114–132. [CrossRef]

72. Dunne, T.; Black, R.D. Partial Area Contributions to Storm Runoff in a Small New England Watershed. *Water Resour. Res.* **1970**, *6*, 1296–1311. [CrossRef]

73. Gallart, F.; Llorens, P.; Latron, J. Studying the role of old agricultural terraces on runoff generation in a small Mediterranean mountainous basin. *J. Hydrol.* **1994**, *159*, 291–303. [CrossRef]

74. Schnabel, S.; Lozano Parra, J.; Gómez-Gutiérrez, A.; Alfonso-Torreño, A. Hydrological dynamics in a small catchment with silvopastoral land use in SW Spain. *Cuad. Investig. Geográfica* **2018**, *44*, 557–580. [CrossRef]

75. Dahlgren, R.A.; Tate, K.W.; Lewis, D.J.; Atwill, E.R.; Harper, J.M.; Allen-Diaz, B.H. Watershed research examines rangeland management effects on water quality. *Calif. Agric.* **2001**, *55*, 64–71. [CrossRef]

76. Licciardello, F.; Barbagallo, S.; Gallart, F. Hydrological and erosional response of a small catchment in Sicily. *J. Hydrol. Hydromech* **2019**, *67*, 201–212. [CrossRef]

77. Pacheco, E.; Farguell, J.; Úbeda, X.; Outeiro, L.; Miguel, A. Runoff and sediment production in a mediterranean basin under two different land uses. *Cuaternario y Geomorfol.* **2011**, *25*, 103–114.

78. Ben Slimane, A.; Raclot, D.; Rebai, H.; Le Bissonnais, Y.; Planchon, O.; Bouksila, F. Combining field monitoring and aerial imagery to evaluate the role of gully erosion in a Mediterranean catchment (Tunisia). *CATENA* **2018**, *170*, 73–83. [CrossRef]

79. Roig-Planasdemunt, M.; Llorens, P.; Latron, J. Seasonal and storm flow dynamics of dissolved organic carbon in a Mediterranean mountain catchment (Vallcebre, eastern Pyrenees). *Hydrol. Sci. J.* **2016**, *62*, 1–14. [CrossRef]

80. Cayuela, C.; Latron, J.; Geris, J.; Llorens, P. Spatio-temporal variability of the isotopic input signal in a partly forested catchment: Implications for hydrograph separation. *Hydrol. Process.* **2019**, *33*, 36–46. [CrossRef]

MDPI
St. Alban-Anlage 66
4052 Basel
Switzerland
Tel. +41 61 683 77 34
Fax +41 61 302 89 18
www.mdpi.com

Water Editorial Office
E-mail: water@mdpi.com
www.mdpi.com/journal/water